Active Motion and Swarming

From Individual to Collective Dynamics

D I S S E R T A T I O N

zur Erlangung des akademischen Grades

doctor rerum naturalium
(Dr. rer. nat.)
im Fach Physik

eingereicht an der
Mathematisch-Naturwissenschaftlichen Fakultät I
Humboldt-Universität zu Berlin

von
Dipl.-Phys. Pawel Piotr Romanczuk
geboren am 08.08.1978 in Gdańsk (Polen)

Präsident der Humboldt-Universität zu Berlin:
Prof. Dr. Jan-Hendrik Olbertz

Dekan der Mathematisch-Naturwissenschaftlichen Fakultät I:
Prof. Dr. Andreas Herrmann

Gutachter:
1. Prof. Dr. Lutz Schimansky-Geier
2. Prof. Dr. Igor M. Sokolov
3. Prof. Dr. Markus Bär

eingereicht am: 16.12.2010
Tag der mündlichen Prüfung: 16.02.2011

Reihe Nichtlineare und Stochastische Physik
herausgegeben von:
Prof. Dr. Lutz Schimansky-Geier
Institut für Physik
Humboldt-Universität zu Berlin
Newtonstr. 15
D-12489 Berlin

email: alsg@physik.hu-berlin.de

Zugl.: Diss., Univ., Humboldt-Universität zu Berlin, 2011

Bibliografische Information der Deutschen Nationalbibliothek

Die Deutsche Nationalbibliothek verzeichnet diese Publikation in der
Deutschen Nationalbibliografie; detaillierte bibliografische Daten sind
im Internet über http://dnb.d-nb.de abrufbar.

ISBN 978-3-8325-2990-1
ISSN 1435-7151

Logos Verlag Berlin
Comeniushof, Gubener Str. 47,
10243 Berlin
Tel.: +49 030 42 85 10 90
Fax: +49 030 42 85 10 92
INTERNET: http://www.logos-verlag.de

Abstract

In this work, we introduce and discuss different models of individual active Brownian motion as well as systems of interacting active particles. The first part of this work deals with individual active Brownian dynamics in one and two spatial dimensions. In one dimension, we analyze models of active Brownian motion with decoupled velocity and heading dynamics.

In the two-dimensional case, a particular focus lies on the distinction of active and passive fluctuations. It is shown that, in general, active velocity fluctuations lead to an increased probability of low speeds and, as a consequence, to characteristic deviations of stationary velocity (speed) distributions in comparison to the case of pure passive fluctuations (e.g. thermal fluctuations). Furthermore, the mean squared displacement for the different models is analyzed and (approximate) analytical expressions for the effective coefficient of spatial diffusion are derived.

In the second part, we extend our scope to systems of interacting active Brownian particles. At first, a kinetic mean field theory of active Brownian motion with velocity alignment interaction is derived in one and two spatial dimensions. The stationary solutions corresponding to a disordered state (no collective motion) and to an ordered state of collective motion are identified for the spatially homogeneous system. Their stability and the corresponding phase transition behavior with respect to noise and coupling strength is discussed.

Further on, a model of (active) Brownian particles interacting via biologically motivated escape and pursuit interactions is introduced. The onset of large scale collective motion is shown for both interaction types and the different impact of escape and pursuit on the global dynamics is analyzed. Based on analysis of collective motion of particle pairs, analytical approximations for the speed of collective motion are obtained which yield the right scaling of mean speed of collective motion also for large systems. In addition, a mean field theory of escape and pursuit and a generalization of the escape and pursuit model for self-propelled particles based on selective attraction and repulsion to approaching or receding individuals are briefly discussed.

Recent experiments demonstrate that the nutritional state of individual locusts does not affect their motion when alone, but has a significant impact on the motion of socially interacting locusts which manifests in larger speed of protein deprived locusts in groups. It is shown that a modified escape and pursuit model which accounts for the observed dynamics of individual locusts is able to explain and reproduce the impact of nutritional condition on the group behavior by associating different nutritional states with different interaction strengths of escape and pursuit. Furthermore, the model allows predictions on the impact of the nutritional state of individuals on the critical density of individuals required for the onset of large-scale collective migration in locusts.

Zusammenfassung

In der vorliegenden Arbeit werden unterschiedliche Modelle der Bewegung einzelner aktiver Brownscher Teilchen sowie gekoppelter aktiver Teilchen eingeführt und untersucht. Der erste Teil der Arbeit behandelt individuelle Brownsche Dynamik in ein und zwei Raumdimensionen. Im eindimensionalen Fall wird aktive Brownsche Bewegung betrachtet, bei der die Geschwindigkeit und Bewegungsrichtung durch unabhängige stochastische Prozesse beschrieben werden. Im zweidimensionalen Fall liegt der Schwerpunkt der Untersuchungen auf der Unterscheidung von aktiven und passiven Fluktuationen. Es wird gezeigt, dass aktive Fluktuationen im allgemeinen zu einer erhöhten Wahrscheinlichkeit geringer Geschwindigkeiten führen, die als Konsequenz im Vergleich mit rein passiven Fluktuationen (z.B.: thermische Fluktuationen) zu charakteristischen Abweichungen der stationären Geschwindigkeitsverteilungen bzw. Geschwindigkeitsbetragsverteilungen führen. Des Weiteren wird die mittlere quadratische Verschiebung der unterschiedlichen Modelle analysiert und (genäherte) analytische Ausdrücke für die entsprechenden effektiven Koeffizienten der räumlichen Diffusion hergeleitet.

Im zweiten Teil wird die Analyse auf Systeme von gekoppelten Teilchen erweitert. Als erstes wird eine kinetische Theorie der aktiven Brownschen Bewegung mit Geschwindigkeitsgleichrichtung für eine und zwei Raumdimensionen hergeleitet. Die stationären Lösungen des räumlich homogenen Falls, die einem ungeordneten Zustand (keine kollektive Bewegung) und einem geordneten Zustand kollektiver Bewegung entsprechen, werden identifiziert. Die Stabilität der Lösungen und die zugehörigen Phasenübergänge im Bezug auf die Änderung der Rausch- und der Kopplungsstärke werden untersucht. Des Weiteren wird ein Modell von aktiven Brownschen Teilchen eingeführt mit biologisch motiviertem "Flucht"- und "Verfolgungs"-Verhalten. Die Entstehung großskaliger kollektiver Bewegung wird für beide Interaktionsarten gezeigt und deren unterschiedlicher Einfluß auf das globale Verhalten des Systems wird untersucht. Basierend auf der Analyse von wechselwirkenden Teilchenpaaren werden analytische Näherungen für die resultierende Geschwindigkeit der kollektiven Bewegung hergeleitet, die ebenfalls die Skalierung der mittleren Geschwindigkeit kollektiver Bewegung in Abhängigkeit der Modellparameter für große Systeme liefert. Darüber hinaus wird ein Mean-Field Ansatz zur Analyse des "Flucht+Verfolgung" Modells herangezogen. Es wird ebenfalls ein generalisiertes Modell für selbstangetriebene Teilchen, basierend auf selektiver Anziehung und Abstoßung in Abhängigkeit von der relativen Geschwindigkeit interagierender Teilchen, eingeführt.

Neueste experimentelle Untersuchungen zeigen, dass in Abwesenheit anderer Artgenossen der Ernährungszustand von Heuschrecken keinen Einfluss auf deren Bewegungverhalten hat. Im Gegensatz hierzu, hat der Ernährungszustand einen signifikanten Einfluss auf die Bewegung sozial interagierender Heuschrecken, was sich in höheren Geschwindigkeiten proteindeprivierter Einzelindividuen innerhalb einer Gruppe äussert. Es wird gezeigt, das ein modifiziertes "Flucht+Verfolgung" Modell, welches die Dynamik der Einzelindividuen realistisch beschreibt, die experimentellen Ergebnisse erklären und reproduzieren kann, unter der Annahme eines Zusammenhangs zwischen dem Ernährungszustand und der Stärke der effektiven sozialen Wechselwirkung. Darüber hinaus erlaubt das Modell Vorhersagen über den Einfluss des Ernährungszustandes von Einzelindividuuen auf die kritische Dichte von Insekten, bei der kollektive Schwarmbewegung einsetzt.

Contents

1. Introduction

One of the greatest triumphs in statistical physics at the beginning of the 20th century was the formulation of the theory of Brownian motion associated with such famous names as Einstein (1905), Langevin (1908) and von Smoluchowski (1906). Since its first publication by Einstein in 1905, the theory of Brownian motion not only assumed a central role in the foundation of thermodynamics and statistical physics, but is still a major interdisciplinary research topic.

In recent decades, there has been an increasing focus on the statistical description of systems far from equilibrium. A whole class of biological and physical systems which may be referred to as *active matter* have been studied theoretically and experimentally. The term "active" refers here to the ability of individual units to move actively by gaining kinetic energy from the environment. Examples of such systems range from the dynamical behavior of individual units such as Brownian motors (Reimann, 2002), motile cells (Friedrich and Julicher, 2007; Selmeczi et al., 2008; Bödeker et al., 2010), macroscopic animals (Kareiva and Shigesada, 1983; Komin et al., 2004) or artifical self-propelled particles (Paxton et al., 2004; Howse et al., 2007) to large coupled ensembles of such units and their large-scale collective dynamics (Vicsek et al., 1995; Chaté et al., 2006; Baskaran and Marchetti, 2009). A major driving force of the *active matter* research are the continuously improving experimental techniques such as for example automated digital tracking (Sokolov et al., 2007; Selmeczi et al., 2008; Ballerini et al., 2008) or the realization of active granular and colloidal systems (see e.g. Paxton et al., 2004; Kudrolli et al., 2008; Tierno et al., 2010; Deseigne et al., 2010).

Despite recent advances in active matter research, there still remain many open questions regarding its universal properties and the aim of this work is to address few of them from a theoretical perspective. An important feature of most *active matter* systems are the non-negligible random fluctuations in the motion of individual active units. This apparent randomness may have different origins, for example, environmental factors or internal fluctuations due to the intrinsic stochasticity of the processes driving individual motion. In animals, they may be also associated with abstract decision processes which govern the direction and/or the speed of individual motion and which may appear as random to an external observer. A simple way to account for such fluctuations, without being able to resolve the underlying mechanisms, is to introduce stochastic forces into the equations of motion of individual units. Thus the general modelling approach is based on the concept of stochastic differential equations[1] (SDE) and the corresponding Fokker-Planck equations for the evolution of the probability densities of the involved (stochastic) variables (Gardiner, 1985; van Kampen, 1992; Risken, 1996).

This work stands in the long tradition of statistical physics and stochastic processes applied to biological systems. In particular, it was strongly influenced by the concept of Active Brownian Particles developed more then a decade ago at the Humboldt Universität

---

[1]In physics, SDEs are also referred to as Langevin equations.

zu Berlin.

The term "Active Brownian Particles" was first introduced by Schimansky-Geier et al. (1995), referring to Brownian particles with the ability to generate a field, which in turn can determine their motion. This first publications was followed by many others, which used the term in the context of Brownian particles with an internal energy depot (see e.g. Schweitzer et al., 1998; Ebeling et al., 1999; Erdmann and Ebeling, 2003; Schweitzer, 2003; Romanczuk et al., 2008).

Although my research was originally inspired by biological systems, the generic theoretical models discussed here may also applied to artificial active systems far from equilibrium. The monograph itself is structured in two parts: In the first part, I discuss models of individual active motion, whereas the second parts analyzes the onset of collective motion in large systems of interacting active particles. In the forthcoming sections, I give a brief (historical) introduction for the two parts and relate the contents of this work to previous publications.

1.1. Individual Dynamics

Although the name of the botanist Robert Brown is associated with the phenomenon of Brownian motion, he was certainly not the first one to observe it. The phenomenon of irregular motion of coal dust particles immersed in a fluid had already been reported by Jan Ingen-Housz in 1784 (Ingen-Housz, 1784). The contribution of Robert Brown was to perform careful observations of the phenomenon for all sorts of organic and inorganic solid particles in 1827 and to conclude that this random motion is not a manifestation of life but a general property of microscopic objects in a fluid (Brown, 1828).

Einstein and von Smoluchowski explained Brownian motion by linking the random motion of microscopic particles to the thermal motion of molecules of the surrounding fluid. It was a major breakthrough for the kinetic theory of heat by demonstrating that Brownian motion can be predicted directly from the kinetic model of thermal equilibrium.

Probably the first experiments on random motion of (living) microorganisms – which consitute clearly a system far from equilibrium – influenced by Einsteins and Smoluchowskis discovery were performed by Przimbam in the second decade of the 20th century (Przibram, 1913, 1917). Przibram showed that the mean squared displacement of the protozoa in water increases linearly in time in analogy to Brownian motion but with a larger diffusion coefficient than predicted by the equilibrium kinetic theory of Brownian motion. Przimbrams work is the first experimental evidence of active Brownian motion. In fact, in his second paper Przimbram's reported increasing diffusion coefficients of rotifiers with their increasing concentration, which appears to be the first report on hydrodynamically interacting active Brownian particles (Przibram, 1917).

Przimbam's work was followed up by Fürth (1920), who based on his own experimental findings, introduced the notion of persistent random walk in the description of the motion of biological agents. Fürth arrived independently at the same result as Ornstein (1919), who considered inertial Brownian motion in thermal equilibrium.

The mathematical description of the apparently random motion of biological agents and the corresponding diffusion processes are fundamental to understanding the ability of individuals to explore their environment and to describing the large-scale dispersal of populations (Okubo and Levin, 2001). Since the first pioneering works of Przimbam and

Fürth, there have been a great number of publications on the theory of random walks and their application to biology and ecology. Examples are the "rediscovery" of persistent random walk as a model for insect dispersal by Kareiva and Shigesada (1983) — more then 60 years after Fürth, or the numerous works devoted to the application of the stochastic telegraph equation (Kac, 1974) to biological dispersion (Othmer et al., 1988; Hadeler, 1996; Hillen and Othmer, 2000). The properties of a corresponding random walk models have important biological implications if applied to animal searching behavior, as they have a major impact on the ability of individuals animals to exploit spatially distributed patches of nutrients (see e.g. Komin et al., 2004; Schimansky-Geier et al., 2005; Bartumeus et al., 2005, 2008).

New possibilities of automated digital tracking of microorganisms allows a quantitative analysis of a large number of individual trajectories. Based on this data stochastic equations of motion for human keratinocytes and fibroblasts (Selmeczi, 2005) and amoebae (*Dictyostelium discoideum*) (Bödeker et al., 2010) have been proposed.

An important contribution from the physics perspective was the derivation of the effective diffusion coefficient of self-propelled particles with continuous angular diffusion by Mikhailov and Meinköhn (1997). Their work was motivated by the observations of active motion of inanimate objects at far-from equilibrium conditions, for example the motion of droplets on a substrate due to unbalanced capillary forces (Santos and Ondarçuhu, 1995).

To my knowledge, Schienbein and Gruler (1993) were the first to describe the velocity dynamics of biological agents by a Langevin equation with a velocity dependent friction function to account for the active motion of human granulocytes. It should be noted that Schienbein and Gruler describe the two-dimensional motion of cells via a Langevin equation for the speed with additive Gaussian white noise. In order to exclude the possibility of negative speed (absolute value of velocity), this approach requires some additional constraint, such as a reflecting boundary condition at zero speed, which is not clearly stated in their original paper.

The Langevin ansatz was further promoted in the framework of Brownian particles with internal energy depot (Schweitzer et al., 1998). The depot model can be reduced to a Langevin equation of active Brownian motion with a (nonlinear) velocity dependent friction function if one assumes fast energy depot relaxation. The dynamics of such individual active Brownian particles has been studied with and without external forces in one and two spatial dimensions (see e.g. Ebeling et al., 1999; Erdmann et al., 2000, 2002; Lindner and Nicola, 2008b,a).

In a number of studies it was shown that such velocity dependent friction functions can be used in order to describe active motion in different physical systems far from equilibrium, such as the motion of dissipative solitons (Liehr et al., 2003) or self-moving oil droplets (Sumino et al., 2005).

Despite the intense interdisciplinary research on active Brownian motion, there is still a lack of theoretical foundations. For example, one issue that has been neglected is the distinction between passive fluctuations (e.g. thermal fluctuations) and stochastic forces which have their origin in the active nature of the system and their different impact on the experimentally accessible observables, such as stationary velocity and speed distributions. Only recently, Peruani and Morelli (2007) have shown how internal (or active) fluctuations may lead to a complex behavior of the mean squared displacement of active particles with multiple crossovers.

In the first part of my work, I introduce and analyze different models of active Brownian motion. In Chapter 2, I start with the one-dimensional motion of active particles with a velocity dependent friction function of the Schienbein-Gruler type. In contrast to the original work by Schienbein and Gruler, the velocity with respect to the direction of motion is considered as a degree of freedom. This different approach does not require the introduction of any artificial boundary conditions in the equations of motion (Langevin equation). Furthermore, in contrast to previous publications on one dimensional active Brownian motion (Lindner and Nicola, 2008b,a), the stochastic dynamics of the velocity and direction of motion of the particle are assumed to be given by independent stochastic processes, which in particular for biological agents appears to be a more realistic ansatz. An analysis and discussion of the Schienbein-Gruler velocity model in two dimensions is given in Chapter 3. Here, a particular focus is the distinction of active and passive fluctuations, and their different impact on stationary velocity and speed distributions and the effective coefficient of spatial diffusion.

In Chapter 5, I turn my attention to different models of active Brownian motion where the speed – and not the velocity – is described by a Langevin equation. Based on the structure of different propulsion functions (negative friction functions), I introduce a generalized propulsion function of active motion. In this general framework, I analyze the behavior of the first two speed moments (mean speed and variance) and discuss the behavior of the mean squared displacement.

Finally, at the end of the first part (Chapter 5), I address the question of optimal speeds of animal motion by discussing simple models of metabolism with speed dependent cost of locomotion.

1.2. Collective Motion

Probably the most prominent (minimal) model of collective motion in statistical physics was introduced by Vicsek et al. (1995). In the Vicsek model individual particles move with a constant speed and at each time step assume the average direction of motion of the particles in their local neighbourhood with some noise added. In the limit of vanishing speed, the Vicsek model reduces to the well known XY-model of a ferromagnet (Kosterlitz and Thouless, 1973; Kosterlitz, 1974). Vicsek *et al.* have shown in their groundbreaking work how an initially disordered state at large noise intensities becomes unstable if the noise is decreased below a critical value and large-scale collective motion emerges via a spontaneous symmetry breaking. In contrast to the XY-model, long-range order may emerge in the Vicsek model due to its non-equilibrium nature (Vicsek et al., 1995; Toner and Tu, 1995, 1998). Subsequent studies of the Vicsek model raised questions about the nature of this phase transition (continuous or discontinuous) and lead to an intense debate on the issue (Grégoire and Chaté, 2004; Nagy et al., 2007; Aldana et al., 2007; Chaté et al., 2008).

A common feature of a majority of models of collective motion is the assumption of a constant speed of individuals, which may have important consequences. As, for example, in swarms of self-propelled particles with constant speed and harmonic attraction but without any velocity alignment no stable collective motion can be observed. This changes if we consider active Brownian particles with nonlinear friction function, such as the Rayleigh-Helmholtz friction. The nonlinear coupling of different velocity components in the presence of a harmonic attraction, leads to stable collective modes of motion such as directed translation or rotation of the swarm (Niwa, 1994; Erdmann et al., 2005; Ebeling and Schimansky-Geier, 2008; Strefler et al., 2008).

Collective motion of self-propelled particles may also emerge via other interactions, such as volume exclusion of elongated particles (Peruani et al., 2006) or inelastic collisions (Grossman et al., 2008). Recently, there has been also an increasing scientific interest in the hydrodynamics interactions of active particle suspensions both from theoretical and experimental perspective (see, e.g., Simha and Ramaswamy, 2002a,b; Erdmann and Ebeling, 2003; Hernandez-Ortiz et al., 2005; Saintillan and Shelley, 2007; Sokolov et al., 2007, 2009; Baskaran and Marchetti, 2009; Rushkin et al., 2010). These interactions provide important insights into the onset of collective motion in bacterial colonies and in systems of interacting artificial self-propelled particles.

In contrast to these "direct" mechanisms for the onset of collective motion mediated by particle contacts or via a surrounding fluid, the interactions of animals such as birds and fish depend on their sensory inputs and complex decision processes of individuals. Probably one of the first individual based models of collective motion of higher animals was introduced by Suzuki and Sakai (1973). This model is also described in detail in Okubo and Levin (2001). In this model, as well as in most models for collective motion in biology and ecology suggested so far, the interaction between individuals consists of three distinct components:

- attraction to other individuals (group cohesion),

- repulsion from others at short distances (collision avoidance),

- alignment of velocities of neighboring individuals.

Different variations of this general scheme have been introduced and analyzed over the years (see e.g. Aoki, 1982; Reynolds, 1987; Huth and Wissel, 1992; Couzin et al., 2002) where the different interactions were implemented either as effective social forces or individual behavioral rules.

Although the different components of the effective social force are reasonable in the biological context and yield patterns of collective motion which resemble qualitatively the patterns observed in natural systems, many fundamental questions about the nature of social interactions in real systems are still open. A major shortcoming of the research in collective motion was the lack of empirical data on the structure and dynamics of real swarms. However, the situation is improving; due to technological advances in digital tracking and data processing, the empirical study of large-scale collective motion in the field has become possible (Ballerini et al., 2008; Lukeman et al., 2010). Furthermore, an increasing number of controlled laboratory experiments is being performed on the collective motion of such different organisms as fish (Abaid and Porfiri, 2010) or insects (Buhl et al., 2006; Bazazi et al., 2010). The resulting data provide the foundation to address the question on actual properties of social interactions in real world swarms and flocks. Recently, for example, Ballerini et al. (2008) have suggested, based on empirical data on starling flocks, that the social interactions depend rather on the topological distance then on the metric distance used in most models.

From the biological point of view, there are different possible advantages for individual animals of living and moving in groups such as protection against predators or increased foraging success due to effective exchange of information within groups (Krause and Ruxton, 2002; Guttal and Couzin, 2010). Recently, empirical evidence has been provided for aggression among conspecifics (cannibalism) being the driving force of collective migration in certain insect species, such as Mormon crickets (*Anabrus simplex*; Simpson et al., 2006) or desert locusts (*Schistocerca gregaria*; Bazazi et al., 2008). In these species, it appears that under the threat of cannibalism the insects perform a forced march which leads to the onset of collective migration as a non-cooperative phenomenon.

In the second part of my work, which deals with collective dynamics, I start by considering a gas of active Brownian particles interacting only via a local velocity-alignment force (Chapter 6). Based on a moment expansion of the probability density, I derive systematically a kinetic mean field theory of the interacting particle gas in one and two spatial dimensions for different velocity dependent friction functions. The solutions of the mean field equations for the spatially homogeneous case are discussed in comparison with numerical simulations.

In Chapter 7, I introduce a model of individuals interacting via escape and pursuit response, which may be associated with cannibalism. It is shown that this kind of interactions gives rise to robust collective motion of individuals with strongly fluctuating velocity (Brownian particles) at high densities irrespective of the model details, whereas at low densities the onset of collective motion depends strongly on the relative strength of escape and pursuit. At the end of the chapter, I discuss briefly a generalized model of selective attraction repulsion interactions for self-propelled particles.

Finally, at the end of my work (Chapter 8), I discuss and analyze recent experimental data provided by S. Bazazi (University of Oxford) on the different impact of nutritional condition on the motion of solitary desert locusts in comparison to locusts in a group. Based on these results, a modified escape and pursuit model accounting in detail for the observed motion of individual locusts, is introduced. The modified model is not only able to reproduce the experimentally observed speed distributions but also allows predictions on the nutrition dependence of the critical density of insects for the onset of collective motion.

Part I.

Theory of Individual Dynamics

2. Schienbein-Gruler Velocity Model in 1D

In this chapter, we introduce and analyze simple models for an active particle in one spatial dimension with fluctuating velocity and direction of motion.

Many organisms have a distinct body axis defining their preferred direction of motion (head-tail axis). This introduces a distinct orientation, which we will refer to as *heading*. Also, for artificial active particles it might be natural to assume a preferred direction of motion based on their propulsion mechanism.

In a strictly one dimensional system the direction of motion is a discrete variable which may take only two discrete values $\{-1, +1\}$ corresponding to the motion in positive and negative x-direction. Thus, in the following, we define a heading $h(t)$ of a particle as its preferred direction of motion in one dimension, which can take the two values $h = -1, +1$ and may in general also be time dependent.

For the simple case that the particle does not change its heading the equations of motion may be written as independent of h:

$$\dot{x}(t) = v(t) \ , \tag{2.1}$$

$$\dot{v}(t) = -\gamma(v)v + \eta(t) \ , \tag{2.2}$$

here the particle mass is set to unity ($m = 1$). The first term in Eq. 2.2 is a velocity dependent friction function $\gamma(v)$ responsible for the deterministic velocity dynamics. For the simple example of linear friction $\gamma(v) = const. > 0$ (Stokes friction) the friction term results always in a deceleration of the particle. An arbitrary velocity dependent friction function may also be negative on finite velocity intervalls and therefore lead to acceleration of the particle. Here we choose the friction term $-\gamma(v)v$ to be a linear function of v:

$$-\gamma(v)v = \alpha\,(v_0 - v)\,, \tag{2.3}$$

with v_0 being the stationary velocity of the particle with respect to its heading. The constant $\alpha = \tau_v^{-1} > 0$ is the inverse relaxation time of the velocity dynamics. This friction term can be considered as a polar version of the model proposed by Schienbein and Gruler (1993). In contrast to the Schienbein-Gruler friction discussed in Erdmann et al. (2000), the above friction term results in a single fixed point of the velocity dynamics. The case of $v_0 \geq 0$, corresponds for a constant heading to a preferred direction of motion in positive x-direction, whereas $v_0 \leq 0$ corresponds to motion in negative x-direction. Thus in this case we may define the heading simply as: $h = v_0/|v_0|$.

For $v_0 \geq 0$ the friction term in Eq. 2.3 is negative for $v > v_0$ (friction) and positive for $v < v_0$ (propulsion) and reduces to Stokes friction for $v_0 = 0$. The velocity potential defined through $-dV(v)/dv = -\gamma(v)v$ is harmonic and centered around v_0:

$$V(v) = \alpha(v - v_0)^2/2. \tag{2.4}$$

Please note that for $v_0 \neq 0$ the potential is not symmetric with respect to $v = 0$.

The second term in Eq. 2.2 is a random force, $\eta(t)$, which is responsible for the velocity fluctuations. Here we consider the fluctuations $\eta(t)$ to be Gaussian white noise with vanishing correlations and zero mean:

$$\eta(t) = \sqrt{2D}\xi(t), \quad \langle \xi(t) \rangle = 0, \quad \langle \xi(t)\xi(t') \rangle = \delta(t - t'). \tag{2.5}$$

In the deterministic case $\eta(t) = 0$ the velocity of the particle relaxes for arbitrary conditions $v(t = 0)$ to the stationary velocity v_0, whereas for finite fluctuation strengths the velocity fluctuates around v_0 with $\tau_v = \alpha^{-1}$ being the velocity correlation time.

The stationary velocity probability density function (PDF) for particles with the same heading, e.g. $h = 1$, can easily be calculated by solving the corresponding stationary Fokker-Planck equation (Risken, 1996) and is given by a Gaussian PDF with mean v_0 and variance $\sigma^2 = D/\alpha$:

$$p_s(\dot{x}) = p_s(v) = N \exp\left(\frac{-\alpha(v - v_0)^2}{2D} \right). \tag{2.6}$$

Here, $\mathcal{N} = \sqrt{\alpha/(2\pi D)}$ is a normalization constant which ensures $\int_\infty^\infty dv p(v) = 1$. Please note that for $D > 0$ the observed direction of motion[1] defined by the sign of $\dot{x}$ might be different to the sign of v_0. Due to velocity fluctuations there exists a finite probability of negative velocities with respect to the preferred direction of motion. Nevertheless, in the stationary case the observed displacement averaged over time (or ensemble) $\langle \dot{x} \rangle$ is given by v_0.

We consider now an ensemble of particles with v_0 either positive or negative. The fraction of particles with $v_0 > 0$, $h = +1$ is q_+, whereas q_- is the fraction particles with $v_0 < 0$, $h = -1$. The stationary PDF of observed velocities $P(\dot{x})$ for such an ensemble of active particles with constant heading is entirely determined by the h-distribution

$$\hat{p}(h) = q_+\delta(h - 1) + q_-\delta(h + 1).$$

The observed velocity PDF of the ensemble $P(\dot{x})$ is given by the superposition of the stationary velocity PDFs for the two direction of motion $p_+(\dot{x}) = p(\dot{x}, h = 1)$ and $p_-(\dot{x}) = (\dot{x}, h = -1)$:

$$P(\dot{x}) = q_+ p_+ + q_- p_-$$
$$= \mathcal{N}\left[q_+ \exp\left(-\frac{\alpha(\dot{x} - v_0)^2}{2D} \right) + q_- \exp\left(-\frac{\alpha(\dot{x} + v_0)^2}{2D} \right) \right] \tag{2.7}$$
$$= \left[1 + q_+ \left(\exp\left(-\frac{2\alpha\dot{x}v_0}{D} \right) - 1 \right) \right] p_s(\dot{x}) \tag{2.8}$$

Here, in order to simplify the expression for $P(\dot{x})$ we have used (2.6) and $q_- = 1 - q_+$.

Our discussion of the model has, up until this point, been confined to the simple case of

---

[1]In the following, we distinguish the velocity v of a particle with respect to its preferred direction of motion from $\dot{x}$, which is the velocity of the particle measured by an external observer. These two quantities may differ depending on the dynamics of the heading coordinate (see Sec. 2.1).

the constant heading. Now we turn to the case where the heading of an individual particle is time dependent $h = h(t)$.

There have been several publications on one-dimensional motion of active Brownian particles, but to our knowledge the corresponding velocity potentials were always assumed to be symmetric with respect to $v = 0$ with two fixed points at $-v_0$ and v_0 separated by an energetic barrier (Lindner, 2007; Lindner and Nicola, 2008b,a).

For such symmetric bistable velocity potentials the changes in the direction of motion are caused naturally by noise-induced transitions over the barrier. As a result the changes in the heading direction are determined by the noise intensity in the velocity dynamics. Since the Schienbein-Gruler velocity potential (2.4) has only one fixed point an alternative mechanism for heading change of active particles has to be introduced. This allows to decouple the heading dynamics from the velocity fluctuations, which is a reasonable generalization of active Brownian motion as both variables can be governed by independent stochastic processes.

In the following, we consider the case, where an active particle might change its heading h independently of its velocity dynamics. We should emphasize that this does not necessarily mean that on the other hand the velocity dynamics is independent from the changes in the heading as we shall see later. We assume that the change of the heading $h(t)$ from ± 1 to ∓ 1 occurs on a much faster time scale than the velocity dynamics and may be considered as instantaneous. The times between heading changes are assumed to be stochastic. Here we choose them to be given by a Poisson process. Thus the heading $h(t)$ is given by the so-called random telegraph process or dichtomous Markov noise (DMN). The temporal evolution of the probabilities for the two heading states $q(h = \pm 1, t) = q_{+/-}(t)$ is given as

$$\dot{q}_+(t) = \kappa_1 q_-(t) - \kappa_2 q_+(t) \tag{2.9}$$
$$\dot{q}_-(t) = -\kappa_1 q_-(t) + \kappa_2 q_+(t). \tag{2.10}$$

Here κ_i are constant transition rates between the two heading states. Furthermore, at any point in time the probability to find the particle in either one in the states is $q_+(t) + q_-(t) = 1$. The stationary probabilities for the two states $(\dot{q}_+ = \dot{q}_- = 0)$ are:

$$q_+ = \frac{\kappa_1}{\kappa_1 + \kappa_2} \quad \text{and} \quad q_- = 1 - q_+ . \tag{2.11}$$

Based on two different scenarios leading to effectively one dimensional motion of active particles we distinguish two different possibilites how the heading changes affect the equations of motion. We will refer to this two different mechanisms as *turning* and *switching*, respectively.

2.1. *Turning* Model

In the *turning* model we assume that the propulsion force of the active Brownian particle is independent from its heading. This can be an active particle with a single constantly working "propulsion" engine. Such a particle moving in an effective one dimensional environment, such as a narrow tube, can only change its preferred direction of motion (heading) by rapid turns as sketched in Figs. 2.1. For simplicity, it is assumed that during

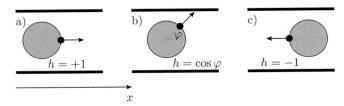

Figure 2.1.: Visualization of the turning model for heading change during effectively one-dimensional motion of active particles. The heading $h = +1$ ($h = -1$) corresponds to the positive (negative) x-direction, which in polar coordinates corresponds to the angle $\varphi = 0$ ($\varphi = \pi$). The particle is assumed to have a single active propulsion engine indicated by the small filled circle. Starting with an initial heading $h = 1$ (a) the particle changes its direction of motion by a rapid turn (b) to continue its motion in the opposite direction $h = -1$ (c).

a turn there is no dissipation of kinetic energy and that the velocity of the particle with respect to the heading does not change. In this case, the temporal evolution of the position $x(t)$ of an individual particle is determined by the following set of (stochastic) differential equations:

$$\dot{x}(t) = v(t)h(t) , \tag{2.12}$$

$$\dot{v}(t) = -\alpha(v - v0) + \eta(t) . \tag{2.13}$$

Please note that $h(t)$ defines only the sign of $\dot{x}$ and does not have any impact on the dynamics of $v(t)$. In particular, the velocity potential $V(v)$ is not affected by $h(t)$.

For simplicity we will restrict the discussion to the unbiased case with $\kappa_1 = \kappa_2 = \kappa$. The average time between two heading changes is κ^{-1}. The correlation time of the heading is given by the correlation time of the DMN: $\tau_h = (2\kappa)^{-1}$. In this symmetric situation there is no distinguished direction of motion. The (average) fraction of particles with $h = \pm 1$ is $q_+ = q_- = 0.5$ and the mean velocity vanishes $\langle v \rangle = 0$.

In Fig. 2.2 we show sample trajectories obtained from numerical simulations of the turning model for two ratios of the correlation times τ_v/τ_h. The velocity dynamics in both subfigures differ in correlation times τ_v, but in both cases the heading change corresponds to sharp changes in the velocity $\dot{x}(t) \rightarrow -\dot{x}(t)$.

From the independence of $h(t)$ and $v(t)$, we infer that the observed stationary velocity PDF of the *turning* model is identical to the PDF obtained for stationary headings in (2.8), with q_+ (q_-) being now the stationary fractions of particles moving to the left and right determined by the turning rates given in (2.11) (see Figs. 2.5 and 2.6).

In the absence of any bias we observe a diffusive spread of an initial spatially localized particle distribution. The process is characterized by the time dependent mean squared displacement $\langle \Delta x^2 \rangle = \langle (x(t) - x(0))^2 \rangle$ (MSD). The MSD averaged over different realization of the process can be expressed in terms of the velocity-velocity correlation function

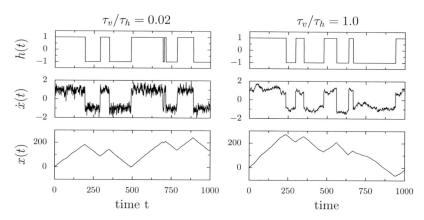

Figure 2.2.: Examples of a single trajectory ($h(t)$, $\dot{x}(t)$ and $x(t)$) of the turning model for two different ratios of the correlation times $\tau_v/\tau_h = 0.02$ (left panels) and $\tau_v/\tau_h = 1.0$ (right panels).

(Ebeling and Sokolov, 2005):

$$\langle \Delta x^2 \rangle(t) = \int_0^t dt' \int_0^t dt'' \langle \dot{x}(t')\dot{x}(t'') \rangle .\tag{2.14}$$

If the stochastic processes governing the evolution of $\dot{x}(t)$ are stationary, the velocity-velocity correlation depends only on the absolute value of the time difference $\tau = |t' - t''|$. Thus we may rewrite the expression for MSD by changing the double integral in (2.14) to the variables τ, t':

$$\langle \Delta x^2 \rangle(t) = 2\int_0^t dt' \int_0^{t'} d\tau \langle \dot{x}(0)\dot{x}(\tau) \rangle .\tag{2.15}$$

For the turning model the velocity at time t is given by the product of $v(t)$ and $h(t)$, as both variables are determined by independent, uncorrelated random processes we rewrite the velocity-velocity correlation as

$$\langle \dot{x}(0)\dot{x}(\tau) \rangle = \langle v(0)h(0)v(\tau)h(\tau) \rangle = \langle v(0)v(\tau) \rangle \langle h(0)h(\tau) \rangle.\tag{2.16}$$

The velocity autocorrelation $\langle v(0)v(\tau) \rangle$ for the Schienbein-Gruler friction in (2.3) decays exponentially with the relaxation time τ_v according to:

$$\langle v(0)v(\tau) \rangle - \langle v \rangle^2 = \left(\langle v^2 \rangle - \langle v \rangle^2 \right) e^{-\frac{\tau}{\tau_v}} .\tag{2.17}$$

13

Thus with the first two moments

$$\langle v \rangle = v_0 \quad \text{and} \quad \langle v^2 \rangle = v_0^2 + D/\alpha$$

and $\tau_v = \alpha^{-1}$ we obtain the corresponding correlation function as

$$\langle v(0)v(\tau) \rangle = v_0^2 + \frac{D}{\alpha} e^{-\alpha\tau}. \tag{2.18}$$

The correlation function of the heading is given by an exponential decay with the relaxation time τ_h:

$$\langle h(0)h(\tau) \rangle = h(0)^2 e^{-\frac{t}{\tau_h}} = e^{-2\kappa t}. \tag{2.19}$$

So for decoupled velocity and heading dynamics we insert (2.18) and (2.19) into 2.15 and obtain:

$$\begin{aligned}
\langle \Delta x^2 \rangle(t) &= 2 \int_0^t dt' \int_0^{t'} d\tau \, v_0^2 e^{-2\kappa\tau} + \frac{D}{\alpha} e^{-(2\kappa+\alpha)\tau} \\
&= \frac{v_0^2}{2\kappa^2} \left(2\kappa t - 1 + e^{-2\kappa t} \right) + \frac{2D}{\alpha(2\kappa+\alpha)^2} \left((2\kappa+\alpha)t - 1 + e^{-(2\kappa+\alpha)\tau} \right). \tag{2.20}
\end{aligned}$$

In the limit of large times $t \to \infty$ the MSD increases linearly in time and the effective coefficient of spatial diffusion reads

$$\begin{aligned}
D_{\text{eff}}^{(1d)} &= \lim_{t\to\infty} \frac{\langle \Delta x^2 \rangle(t)}{2t} = \frac{v_0^2}{2\kappa} + \frac{D}{\alpha(2\kappa+\alpha)} \\
&= \frac{1}{2\kappa} \left(v_0^2 + \frac{D}{\alpha(1+\frac{\alpha}{2\kappa})} \right) \tag{2.21}
\end{aligned}$$

The first term in Eq. (2.21) describes the diffusion due active motion with a mean velocity v_0 in random directions, whereas the second term originates from the velocity fluctuations.

For large α or small κ the contribution from velocity fluctuations to $D_{\text{eff}}^{(1d)}$ may be neglected and the effective diffusion coefficient can be approximated as $D_{\text{eff}}^{(1d)} = v_0^2/(2\kappa)$. In the limit $D \to \infty$ for $2\kappa \ll \alpha$, where $(2\kappa+\alpha) \approx \alpha$, the effective diffusion coefficient is approximately given by the diffusion coefficient of ordinary Brownian motion $D_{\text{eff}}^{(1d)} \to D/\alpha^2$, whereas in the limit of $\kappa \to \infty$ ($\tau_h \to 0$) the effective diffusion coefficient vanishes for finite noise intensities D.

The comparison of the effective diffusion coefficient obtained from numerical simulation of the *turning* model confirm the analytical result in Eq. (2.20) as shown in Fig. 2.7.

2.2. *Switching* Model

In the *switching* model we assume the particle to have two different propulsion engines responsible for motion in either positive or negative direction along the x-axis. At any given time only one of the engines is active. A heading change corresponds to an instantaneous switching of the active engine as shown in Fig. 2.3. The equations of motion in this model

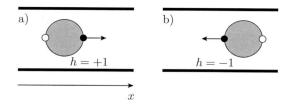

Figure 2.3.: Visualization of the switching model for heading change during one-dimensional motion of active particles. The heading $h = +1$ ($h = -1$) corresponds to the positive (negative) x-direction. The particle is assumed to have two opposite propulsion engines, which propell the particle in different directions. At a certain time only one engine is active (filled small circle) whereas the other one is inactive (empty small circle). A particle switches its heading from $h = 1$ (a) to $h = -1$ (b) by an instantaneous switching of the active engine.

read

$$\dot{x}(t) = v(t) \tag{2.22}$$
$$\dot{v}(t) = -\alpha(v - v_0 h(t)) + \eta(t) \ . \tag{2.23}$$

with $\eta(t)$ given in Eq. 2.5.

For a constant heading $h(t) = 1$ Eq. (2.23) reduces to Eq. (2.13). Thus the stationary PDF $p_s(v)$ in this case is again given by (2.6).

In contrast to the turning model, a change of heading (switching of the engines) directly enters the Langevin equation for the velocity dynamics (2.23). The switching corresponds to a sudden change of the effective potential $V(v)$ from a parabolic potential centered at $\pm v_0$ to one centered at $\mp v_0$. The velocity equation can be reformulated to

$$\dot{v}(t) = -\alpha v + \eta_h(t) + \eta(t), \tag{2.24}$$

which describes a Brownian particle with linear friction subjected to two different noise sources: Gaussian white noise $\eta(t)$ and dichotomous Markov noise $\eta_h(t) = \alpha v_0 h(t)$. The stationary PDFs of such a system for different potential and in different limiting cases have been studied before by Dybiec and Schimansky-Geier (2007).

Examples of trajectories in the switching model for two different ratios of the correlation times τ_h/τ_v are shown in Fig. 2.4. For $\tau_v/\tau_h \ll 1$ the sample trajectory is similar to turning model (left panels, Fig. 2.4), whereas for $\tau_v/\tau_h = 1$ the dynamics clearly differ (right panels, Fig. 2.4). Due to slow relaxation the velocity $v(t)$ does not follow immediately the heading changes. For example, at $t \approx 400$ the heading changes for a short period of time from -1 to $+1$, but $v(t)$ barely reacts and is still close to the original value at the following reverse heading change.

In the limit $\kappa/\alpha \ll 1$, where the mean time between two heading changes given by κ^{-1} is much longer than the relaxation time of the velocity α^{-1}, the transient velocity dynamics

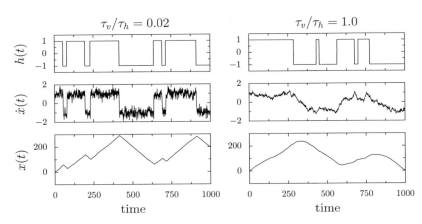

Figure 2.4.: Examples of a single trajectory $(h(t), \dot{x}(t)$ and $x(t))$ of the switching model for two different ratios of the correlation times $\tau_v/\tau_h = 0.02$ (left panels) and $\tau_v/\tau_h = 1.0$ (right panels).

after a heading switch can be neglected and the velocity PDF in both directions may be approximated by the stationary PDF (2.8). In this limit, the behavior of the switching model converges towards the turning model.

In the opposite limit of extremely fast switching with respect to the velocity relaxation time the driving term in the SG friction function in (2.23) self-averages to zero: $\langle v_0 h(t) \rangle = 0$. For $\kappa/\alpha \gg 1$ the switching model converges towards ordinary Brownian motion with constant linear friction coefficient α. Therefore, for increasing κ we expect the switching model to show a crossover from the turning model (independent velocity/heading dynamics) towards ordinary Brownian motion at intermediate values where $\kappa \approx \alpha$. Finally in the limit of large noise intensities D the stochastic propulsion term may be neglected and we again observe ordinary Brownian motion.

The above predictions are confirmed by numerical simulations of the switching model. For small κ ($\ll \alpha$) the velocity probability density coincides with the result obtained in (2.6) (see Fig. 2.5a), whereas for large κ ($\gg \alpha$) it converges towards the Gaussian PDF of ordinary Brownian motion (see Fig. 2.5c). For moderate noise intensities D and $\tau_v \approx \tau_h$ the PDF of the switching model differs from the analytical solutions (see Fig. 2.6a) but the differences vanish with increasing D (see Fig. 2.6c) as both probability densities approach the case of ordinary Brownian motion.

The numerical results for the effective diffusion coefficient $D_{\text{eff}}^{(1d)}$ of the switching model coincide for low κ ($\ll \alpha$) with the analytical result in Eq. (2.21) but converge to D/α^2 for large κ as shown in Fig 2.7a). $D_{\text{eff}}^{(1d)}$ shows similar behavior versus the noise intensity D.

In summary, we have discussed active Brownian motion in one spatial dimension. Two different models of heading dynamics decoupled from velocity fluctuations were introduced and analyzed. For the *turning* model we derived exact expression for the stationary ve-

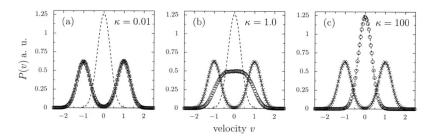

Figure 2.5.: Stationary velocity PDFs for different heading change frequencies κ: (a) 0.01, (b) 1.0, (c) 100.0. Comparison of results obtained from Langevin simulations of the turning model (crosses) and the switching model (circles) to Eq. 2.8 (solid line) and stationary Gaussian PDF of ordinary Brownian motion (dashed line) (other parameters values: $v_0 = 1.0$, $D = 0.1$, $a = 1.0$).

locity PDF and the effective coefficient of diffusion. The *switching* model shows for small switching rates κ similar behavior as the *turning* and reduces to ordinary Brownian motion in the limit of large κ.

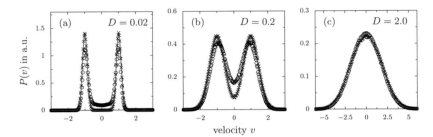

Figure 2.6.: Stationary velocity PDFs for for different noise intensities D: (a) 0.02, (b) 0.2, (c) 2.0. Comparison of results obtained from Langevin simulations of the turning model (crosses) and the switching model (circles) to Eq. 2.8 (solid line) (other parameters values: $v_0 = 1.0$, $\kappa = 0.1$, $a = 1.0$).

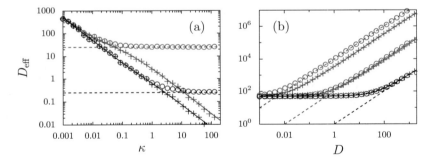

Figure 2.7.: Effective diffusion coefficient $D_{\text{eff}}^{(1d)}$ obtained from simulations of the *turning* model (crosses) and *switching* model (circles) in comparison to results of Eq. (2.21) (solid lines) and the diffusion coefficient of ordinary Brownian motion $D_{\text{eff}}^{(1d)} = D/\alpha^2$ (dashed lines). **(a)** $D_{\text{eff}}^{(1d)}$ versus heading change frequency κ for different $\alpha = 0.2$ (red) and $\alpha = 2.0$ (black) ($D = 1.0$, $v_0 = 1.0$). **(b)** $D_{\text{eff}}^{(1d)}$ versus noise intensity D for $\alpha = 0.01$ (red) $\alpha = 0.1$ (blue) and $\alpha = 1.0$ (black) ($\kappa = 0.01$, $v_0 = 1.0$)

3. Schienbein-Gruler Velocity Model in 2D

In this chapter, we consider the general dynamics of an individual active Brownian point particle with the Schienbein-Gruler friction introduced in Chapter 2 in two spatial dimensions ($d = 2$). The general equation of motion of an active Brownian particle can be written in Cartesian coordinates as

$$\dot{\mathbf{r}} = \mathbf{v} \tag{3.1}$$

$$m\dot{\mathbf{v}} = -\gamma(\mathbf{v})\mathbf{v} + \mathbf{F}_{\text{ext}} + \boldsymbol{\eta}(t) \ . \tag{3.2}$$

Here, $\mathbf{r} = (x, y)$ is the position vector of the particle and $\mathbf{v} = (v_x, v_y)$ is the velocity vector. We set the mass of the particle m without loss of generality to $m = 1$, which corresponds to a rescaling of all forces acting on the particle by m. The first two terms on the right hand side of Eq. 3.2 describe the deterministic evolution of the velocity vector. The first term is again a velocity dependent friction function $\gamma(\mathbf{v})$. The second term $\mathbf{F}_{\text{ext}} = \mathbf{F}_{\text{ext}}(\mathbf{r}, \mathbf{v}, t)$ accounts for possible external forces acting on the particle which in the general case may depend on time as well as on the position and velocity of the particle. The last term is a random force $\boldsymbol{\eta}(t)$ accounting for the randomness in the motion of individual particles.

Before discussing $\boldsymbol{\eta}(t)$ in detail, let us consider first an alternative formulation of the above equations of motion.

In the previous section, we introduced the heading of an active particle as its preferred direction of motion. The concept is now transferred to the two-dimensions case. The turning model (Fig. 2.1), initially formulated as a model for heading changes in an effectively one-dimensional situation, can be naturally extended to two-dimensional motion. In two dimensions, the heading is given by a unit vector $\mathbf{e}_h$ with $|\mathbf{e}_h| = 1$ defining the orientation of the particle. The heading is fully determined by the heading angle $h \equiv \varphi$ defining its direction with respect to the x-axis:

$$\mathbf{e}_h(t) = (\cos\varphi(t), \sin\varphi(t)). \tag{3.3}$$

The evolution of the position of the active particle is given as a product of its velocity with respect to the heading $v(t)$ and $\mathbf{e}_h$ in analogy to the turning model in the one-dimensional case:

$$\dot{\mathbf{r}}(t) = \mathbf{v}(t) = v(t)\mathbf{e}_h(t). \tag{3.4}$$

For the Schienbein-Gruler friction the first term in Eq. (3.2) reads:

$$\gamma(\mathbf{v})\mathbf{v} = \gamma(v)v\mathbf{e}_h = \alpha\,(v - v_0)\,\mathbf{e}_h. \tag{3.5}$$

Here, v_0 is the stationary velocity of the particle with respect to its heading, which in two dimensions may be assumed as $v_0 \geq 0$ without loss of generality. The dynamics of the velocity with respect to the heading are identical to the one-dimensional case as discussed in Chapter 2.

A more convenient representation of the equations of motion can be obtained from the transformation of the velocity vector into the coordinates given by the velocity with repect to the heading $v = \mathbf{v}\mathbf{e}_h$ and the orientation angle φ. The temporal evolution of the velocity vector in the new coordinates reads

$$\dot{\mathbf{v}} = \dot{v}\mathbf{e}_h + v\dot{\varphi}\mathbf{e}_\varphi, \tag{3.6}$$

where $\mathbf{e}_\varphi = (-\sin\varphi(t), \cos\varphi(t))$ is the unit vector in the angular direction, perpedicular to the heading direction ($\mathbf{e}_h\mathbf{e}_\varphi = 0$). Multiplying Eq. 3.2 with $\mathbf{e}_h$ and $\mathbf{e}_\varphi$, respectively, yields two Langevin equations for the evolution of the velocity v and the orientation φ:

$$\dot{v} = -\gamma(v)v + (\mathbf{F}_{\text{ext}} + \boldsymbol{\eta})\,\mathbf{e}_h \tag{3.7a}$$

$$\dot{\varphi} = \frac{1}{v}\,(\mathbf{F}_{\text{ext}} + \boldsymbol{\eta})\,\mathbf{e}_\varphi. \tag{3.7b}$$

Please note that the heading dynamics diverge for $v = 0$. This is due to the fact that we are considering point-like particles and the polar angle φ cannot be defined for a point particle with vanishing velocity. This divergence can be eliminated through a consideration of finite size particles, as will be done in Chapter 4 or by defining a cut-off condition as discussed further below (Sec. 3.2.1).

We should emphasize the difference of the velocity of a particle with respect to its heading v and its speed $|v| = |\mathbf{v}|$. The velocity $v = \mathbf{v}\mathbf{e}_h$ might take also negative values, corresponding to a backward motion of the particle with respect to its heading, whereas the speed given by the absolute value of the velocity is always positive. Thus the velocity-heading coordinates ($v\varphi$-coordinates), have to be distinguished from classical polar coordinates ($|v|,\phi$). The $v\varphi$-coordinate frame is closely related to the Frenet-Serret frame (FS) with the tangential vector $\mathbf{t}$ and normal vector $\mathbf{n}$. For positive velocity $v > 0$ the heading unit vector $\mathbf{e}_h$ equals $\mathbf{t} = \mathbf{v}/|\mathbf{v}|$, whereas $\mathbf{e}_\varphi$ is connected to $\mathbf{n}$. In contrast to the FS frame the curvature K defined through the change of $\mathbf{e}_h$ along the trajectory l:

$$K\mathbf{e}_\varphi = \frac{d\mathbf{e}_h}{dl}, \tag{3.8}$$

may be either positive or negative. Thus, depending on the sign of K, $\mathbf{e}_\varphi$ is either parallel or antiparallel to $\mathbf{n}$. The absolute value of the curvature $|K|$ at a point $l(t_0)$ is the inverse radius $1/R$ of the osculating circle having the same curvature as the trajectory at that point (Fig. 3.1a).

We will focus here on the unbiased motion of individual particles and on the corresponding undirected, purely diffusive spread of active particles in a homogenoues environment with $\mathbf{F}_{\text{ext}} = 0$, but we include the external force term in the formulation of the Fokker-Planck equation in order to account for the general case.

After reformulating the equations of motion in $v\varphi$-coordinates, we proceed with the discussion of the random force $\boldsymbol{\eta}$ in Eq. 3.2. We distinguish two different types of fluctuations, which we will refer to as *passive* (or external) and *active* (or internal) fluctuations, respectively. Passive fluctuations have their origin in a fluctuating environment in which the particle moves. In a homogenoues environment the passive random force $\boldsymbol{\eta}_{\text{p}}$ is independent on the direction of motion of the particle. The classical example of passive

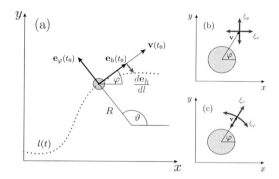

Figure 3.1.: **(a)** Schematic visualization of motion in a plane with unit vectors $\mathbf{e}_h(t)$, $\mathbf{e}_\varphi(t)$ (thick arrows), velocity vector $\mathbf{v}(t)$ (thin arrow) and $d\mathbf{e}_h/dl$ (thick gray arrow). The trajectory of the particle $l(t)$ is indicated by the dotted line. **(b)** Visualization passive fluctuations ξ_x and ξ_y (thick arrows). **(c)** Visualization of active fluctuation ξ_v and ξ_φ (thick arrows).

fluctuations is ordinary Brownian motion in thermal equilibrium, where the stochastic force is associated with random collisions of the molecules of the surrounding fluid with the Brownian particle.

We introduce the passive fluctuations in the same way as in ordinary Brownian motion in two dimensions, as a random noise vector with the components of the vector given by two uncorellated, Gaussian white noise terms with the same noise intensity D. The noise vector reads:

$$\boldsymbol{\eta}_p = \sqrt{2D}\boldsymbol{\xi}(t) = \sqrt{2D}\left(\xi_x(t)\mathbf{e}_x + \xi_y(t)\mathbf{e}_y\right). \tag{3.9}$$

Here $\mathbf{e}_i$ are Cartesian unit vectors and ξ_i are δ-correlated, normally distributed random variables with zero mean:

$$\langle\xi_i(t)\rangle = 0 , \quad \langle\xi_i(t)\xi_j(t')\rangle = \delta_{ij}\delta(t - t'). \tag{3.10}$$

Active fluctuations on the other hand are a pure far-from equilibrium phenomenon and are relevant in the motion of biological agents or self-propelled particles. The origin of these fluctuations can be, for example, variations in the propulsion of chemically powered colloids (Paxton et al., 2004; Howse et al., 2007; Ruckner and Kapral, 2007), complex intracellular processes in cell motility (Selmeczi et al., 2008; Bödeker et al., 2010) or unresolved internal decision processes in animals (Niwa, 1994; Komin et al., 2004; Bazazi et al., 2010). For example, for a macroscopic animal in a homogeneous environment the fluctuations due the environment can be assumed as negligible. Nevertheless, to an external observer the motion of an animal may appear random with respect to its velocity and heading. The apparent randomness of the motion stems from internal descisions of the biological agent to change its direction of motion φ and/or its velocity v. Without being able to resolve

21

the internal descision process, we might describe the changes in the direction and velocity by fluctuations. In this context, we assume the fluctuations in the direction of motion and in the velocity of as independent stochastic processes, with different statistical properties. Here we make a simple ansatz for the active fluctuations, as independent Gaussian white noise in the direction of motion $\mathbf{e}_h$ and in the angular direction $\mathbf{e}_\varphi$:

$$\eta_a = \sqrt{2D_v}\xi_v(t)\mathbf{e}_v + \sqrt{2D_\varphi}\xi_\varphi(t)\mathbf{e}_\varphi \tag{3.11}$$

with different noise intensities of angular and velocity noise: $D_\varphi \neq D_v$. This ansatz allows a direct comparison of passive and active fluctuations. Furthermore, Gaussian noise can be assumed as an approximation to many different stochastic processes due to the center limit theorem (Gardiner, 1985).

The assumption of uncorrelated fluctuations in v and φ is supported by recent experiments on motile cells, which show vanishing crosscorrelations between fluctuations parallel to the direction of motion (velocity fluctuations) and fluctuations perpendicular to the direction of motion (angular fluctuations) (Selmeczi et al., 2008).

3.1. Stationary Velocity and Speed Probability Densities

3.1.1. Only Passive Fluctuations

The equations of motion for passive fluctuations in $v\varphi$-coordinates read

$$\dot{v} = -\gamma(v)v + F_v + \sqrt{2D}(\xi_x(t)\cos\varphi + \xi_y(t)\cos\varphi) \tag{3.12a}$$

$$\dot{\varphi} = \frac{1}{v}\left(F_\varphi + \sqrt{2D}\left(-\xi_x(t)\sin\varphi + \xi_y(t)\cos\varphi\right)\right) \tag{3.12b}$$

with $F_v = \mathbf{F}\mathbf{e}_h$ and $F_\varphi = \mathbf{F}\mathbf{e}_\varphi$ being the components of the external force acting on the velocity and the angle, respectively.

The above Langevin equations have multiplicative noise terms, which have to be taken into account in the derivation of the corresponding Fokker-Planck equation. Here we use the Stratonovich interpretation of the stochastic integral and the coefficients of the Kramers-Moyal expansion can be calculated to

$$K_v^{(1)} = -\gamma(v)v + F_v + \frac{D}{v}\,, \qquad\qquad K_\varphi^{(1)} = \frac{F_\varphi}{v}$$

$$K_{vv}^{(2)} = D\,, \qquad\qquad K_{\varphi\varphi}^{(2)} = \frac{D}{v^2}\,,$$

$$K_{v\varphi}^{(2)} = K_{\varphi v}^{(2)} = 0\,. \tag{3.13}$$

The corresponding Fokker-Planck equation for the PDF $p(v,\varphi)$ reads

$$\frac{\partial p(v,\varphi)}{\partial t} = -\frac{\partial}{\partial v}\left\{\left(-\gamma(v)v + \frac{D}{v} + F_v\right)p - D\frac{\partial p}{\partial v}\right\} - \frac{\partial}{\partial \varphi}\left\{\frac{F_\varphi}{v} - \frac{D}{v^2}\frac{\partial p}{\partial \varphi}\right\}. \tag{3.14}$$

The above equation simplifies in the absence of external forces $\mathbf{F}_{\text{ext}} = 0$ to

$$\frac{\partial p(v,\varphi)}{\partial t} = -\frac{\partial}{\partial v}\left\{\left(-\gamma(v)v + \frac{D}{v}\right)p\right\} + D\frac{\partial^2 p}{\partial v^2} + \frac{D}{v^2}\frac{\partial^2 p}{\partial \varphi^2}. \tag{3.15}$$

Please note that the angular diffusion expressed by the last term of the above equation depends directly on $v(t)$. In the case $\mathbf{F}_{\text{ext}} = 0$, there is no distinguished angular direction. The stationary PDF with respect to φ is homogeneous and we may write $p_p(v,\varphi) = p_p(v)/(2\pi)$, where $p_p(v)$ has to fulfill the following equation:

$$0 = -\frac{\partial}{\partial v}\left\{\left(-\gamma(v)v + \frac{D}{v}\right)p_p(v) - D\frac{\partial}{\partial v}p_p(v)\right\}. \tag{3.16}$$

By inserting the Schienbein-Gruler friction (3.5) and solving the above equation we obtain the stationary velocity distribution for $\mathbf{F}_{\text{ext}} = 0$ as

$$p_p(v) = \mathcal{N}_p|v|e^{-\frac{\alpha(v-v_0)^2}{2D}}, \tag{3.17}$$

with $v \in \mathbb{R}$ and the normalization factor

$$\mathcal{N}_p = \frac{\alpha}{2De^{-\frac{\alpha v_0^2}{2D}} + v_0\sqrt{2\pi\alpha D}\,\text{Erf}\left(\sqrt{\frac{\alpha v_0^2}{2D}}\right)}.$$

This result is confirmed by the velocity distribution obtained from numerical simulations of the Schienbein-Gruler model in two dimensions (SG2d-model) with passive fluctuations as shown in Fig. 3.2. At vanishing noise intensities $D/\alpha \to 0$ the distribution converges towards a δ-peak at v_0. With increasing noise intensity it is approximately given by a narrow Gaussian around v_0 (Fig. 3.2a). With a further increase in D clear deviations from the Gaussian distribution become evident by the appearance of a second maximum of the probability density at negative velocities, caused by backwards motion of the particle with respect to its heading (Fig. 3.2b). The external noise leads to a vanishing probability density at $v = 0$. Finally, in the limit $D/\alpha \to \infty$ the self-propulsion may be neglected and the dynamics converges towards Brownian motion with Stokes friction $-\alpha v$. The velocity PDF becomes symmetric and corresponds to the Rayleigh PDF mirrored along $v = 0$ (Fig. 3.2c).

For arbitrary friction functions and fluctuation types, the speed PDF $\tilde{p}(|v|)$ can be derived from $p(v)$ by taking into account that on average for each polar angle ϕ half of particles head in the direction of ϕ: $\mathbf{e}_{v,+} = (\cos\phi, \sin\phi)$, whereas the other half have the opposite heading $\mathbf{e}_{v,-} = (\cos(\phi+\pi), \sin(\phi+\pi))$. Thus $\tilde{p}(|v|)$ is given by a superposition of the two contributions corresponding to forward and backward motion. For the Schienbein-Gruler friction with only passive fluctuations we obtain:

$$\tilde{p}_p(|v|) = \tilde{\mathcal{N}}_p|v|\left[e^{-\frac{\alpha(|v|-v_0)^2}{2D}} + e^{-\frac{\alpha(|v|+v_0)^2}{2D}}\right] \tag{3.18}$$

$$= \tilde{\mathcal{N}}_p|v|e^{-\frac{\alpha(|v|-v_0)^2}{2D}}\left[1 + e^{-\frac{2\alpha v_0|v|}{D}}\right]. \tag{3.19}$$

In analogy to $p_p(v)$ for $D/\alpha \to \infty$ ($v_0 \to 0$) the distribution converges to the Rayleigh

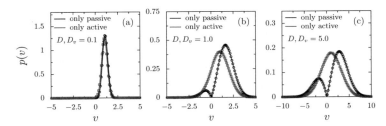

Figure 3.2.: Stationary velocity PDF $p(v)$ for SG-model $v_0 = 1$ under the influence of active (red) and passive (black) fluctuations for different (velocity) noise intensities $D, D_v = 0.1$ (a), 1.0 (b), 5.0 (c). The solid lines show the analytical results for passive fluctuations from Eq. 3.17 and active fluctuations from Eq. 3.25, whereas the symbols show results obtained from numerical simulation of Eqs. 3.7.

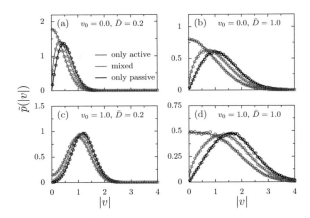

Figure 3.3.: Stationary speed PDFs $\tilde{p}(|v|)$ for Stokes friction $v_0 = 0$ (a,b) and self-propelled Brownian motion $v_0 = 1.0$ (c,d) under the influence of active and passive fluctuations for different total velocity fluctuation strengths $\bar{D} = (D + D_v)/\alpha$. Three different cases are shown: only active fluctuation with $D = 0$, $\bar{D} = D_v/\alpha$ (red), only passive fluctuations $\bar{D} = D/\alpha$, $D_v = 0$ (black) and the mixed case $D_v = D$ (blue); (a,c) $\bar{D} = 0.2$; (b,d) $\bar{D} = 1.0$; The solid lines show the analytical results from Eq. (3.33) whereas the symbols show results obtained from numerical simulation of Eqs. 3.7.

PDF of ordinary Brownian motion

$$\tilde{p}_p(|v|) = \tilde{\mathcal{N}}_p |v| e^{-\frac{\alpha |v|^2}{2D}},$$ (3.20)

whereas for $D/\alpha \to 0$ the limiting probability distribution is a δ-distribution at v_0. The speed PDF vanishes for $|v| = 0$ as a consequence of successive random "kicks" independent of the heading of the particle which drive the particle speed away from $|v| = 0$ (Fig. 3.3).

Finally, we obtain the stationary PDF in Cartesian coordinates $P_p(v_x, v_y)$ from the stationary PDF in polar coordinates $\tilde{p}_p(|v|, \phi) = \tilde{p}_p(|v|)/(2\pi)$ by corresponding coordinate transformation to:

$$P_p(v_x, v_y) = \tilde{\mathcal{N}} e^{-\frac{\alpha \left(\sqrt{v_x^2 + v_y^2} - v_0\right)^2}{2D}} \left[1 + e^{-\frac{2\alpha v_0 \sqrt{v_x^2 + v_y^2}}{D}}\right].$$ (3.21)

Examples of $P_s(v_x, v_y)$ are shown Fig. 3.4.

3.1.2. Only Active Fluctuations

The equations of motion for active fluctuations defined in (3.11) assume a simple form in the $v\varphi$-coordinates:

$$\dot{v} = -\gamma(v)v + F_v + \sqrt{2D_v}\xi_v,$$ (3.22a)

$$\dot{\varphi} = \frac{1}{v}\left(F_\varphi + \sqrt{2D_\varphi}\xi_\varphi\right).$$ (3.22b)

The corresponding Fokker-Planck equation reads

$$\frac{\partial p(v, \varphi)}{\partial t} = -\frac{\partial}{\partial v}\left\{\left(-\gamma(v)v + F_v\right)p - D_v\frac{\partial p}{\partial v}\right\} - \frac{\partial}{\partial \varphi}\left\{\frac{F_\varphi}{v} - \frac{D_\varphi}{v^2}\frac{\partial p}{\partial \varphi}\right\}$$ (3.23)

and simplifies for $\mathbf{F}_{\text{ext}} = 0$ to

$$\frac{\partial p(v, \varphi)}{\partial t} = -\frac{\partial}{\partial v}\left(-\gamma(v)vp\right) + D_v\frac{\partial^2 p}{\partial v^2} + \frac{D_\varphi}{v^2}\frac{\partial^2 p}{\partial \varphi^2}.$$ (3.24)

The corresponding stationary velocity PDF is Gaussian as in the one-dimensional case:

$$p_a(v) = \mathcal{N}_a e^{-\frac{\alpha(v - v_0)^2}{2D_v}}$$ (3.25)

with the normalization factor $\mathcal{N}_a = (\alpha/(2\pi D_v)^{\frac{1}{2}}$.

In the absence of external forces ($\mathbf{F}_{\text{ext}} = 0$), there is no preferred direction of motion and the stationary $v\varphi$-distribution $p_a(v, \varphi) = p_a(v)/(2\pi)$ is independent of φ.

The corresponding speed distribution, which takes into account forward and backward motion with respect to the heading reads

$$p_a(|v|) = \mathcal{N}_a e^{-\frac{\alpha(|v| - v_0)^2}{2D_v}} \left[1 + e^{-\frac{2\alpha|v|v_0}{D_v}}\right].$$ (3.26)

Please note the not-vanishing probability density at $|v| = 0$ and the corresponding

25

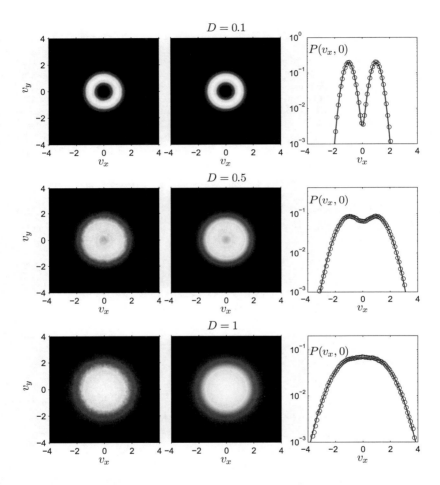

Figure 3.4.: Stationary velocity PDF $P_s(v_x, v_y)$ of the SG2d-model with passive fluctuations for different noise strengths: $D = 0.1$ (top), $D = 0.5$ (center), $D = 1.0$ (bottom). **Left column:** Results obtained from Langevin simulations; **Central column:** Analytical solution given in Eq. 3.21; **Right column:** One dimensional cross-sections $P_s(v_x, 0)$ comparing analytical solutions (solid lines) with numerics (symbols); Other parameters $\alpha = 1.0, v_0 = 1.0$.

absence of the linear increase of $\tilde{p}_a(|v|)$ for small $|v|$ as shown in Fig. 3.3. This atypical behavior indicates that there is no limit where $\tilde{p}_a(|v|)$ converges towards a Rayleigh PDF (Eq. 3.20). This behavior is characteristic for purely active fluctuations where the velocity noise acts along the direction of motion and is independent on the particular choice of the friction function. For $v_0 = 0$ (Stokes friction) the speed distribution simplifies to a half-Gaussian with a maximum at $|v| = 0$:

$$p_a(|v|) = \mathcal{N}_a e^{-\frac{\alpha|v|^2}{2D_v}} . \tag{3.27}$$

The probability distribution in the Cartesian velocity coordinates v_x, v_y can be directly obtained through a coordinate transformation to

$$\tilde{P}(v_x, v_y) = \frac{\mathcal{N}_a}{2\pi} \frac{1}{\sqrt{v_x^2 + v_y^2}} \left[1 + e^{-\frac{2\alpha\sqrt{v_x^2+v_y^2}s_0}{D_v}} \right] e^{-\frac{\alpha(\sqrt{v_x^2+v_y^2}-s_0)^2}{2D_v}} . \tag{3.28}$$

Please note that the velocity distribution in Cartesian coordinates diverges for $v_x = 0$, $v_y = 0$ and, as a consequence, exhibits a sharp peak close to the origin as shown in Fig 3.5.

3.1.3. Mixed Case

In general, the motion of a small particle will be influenced by both fluctuations types $\boldsymbol{\eta}(t) = \boldsymbol{\eta}_a(t) + \boldsymbol{\eta}_p(t)$. Using Stratonovich interpretation the corresponding Fokker-Planck equation can be derived to:

$$\frac{\partial p(v, \varphi)}{\partial t} = -\frac{\partial}{\partial v} \left[\left(-\gamma(v)v + \frac{D}{v} \right) p \right]$$
$$+ (D_v + D)\frac{\partial^2 p}{\partial v^2} + \frac{D_\varphi + D}{v^2}\frac{\partial^2 p}{\partial \varphi^2} . \tag{3.29}$$

The stationary PDF with respect to the angle is again uniform $q(\varphi) = 1/(2\pi)$, whereas the stationary velocity PDF reads:

$$\tilde{p}_m(v) = \mathcal{N}_m |v|^{\frac{D}{D+D_v}} e^{-\frac{\alpha(v-v_0)^2}{2(D+D_v)}} . \tag{3.30}$$

The inverse normalization factor reads

$$\mathcal{N}_m^{-1} = 2^{-A} \left(\frac{D + D_v}{\alpha} \right)^{B+\frac{1}{2}} \Gamma(1 - A) \, _1F_1(-B, \frac{1}{2}, -C)$$
$$+ 2^B v_0 \left(\frac{D + D_v}{\alpha} \right)^{B} \Gamma(1 + B) \, _1F_1(A, \frac{3}{2}, -C)$$
$$+ \frac{2^{B-1} v_0}{\sqrt{\pi}} e^{-C} \left(\frac{D + D_v}{\alpha} \right)^{B} \Gamma(1 + B)\Gamma(1 - A) \, U(1 + B, \frac{3}{2}, C) \tag{3.31}$$

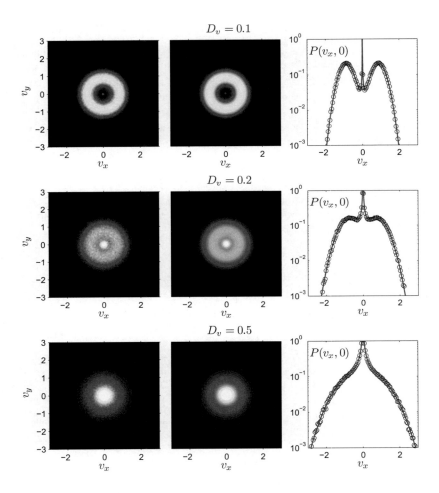

Figure 3.5.: Stationary velocity distributions $P_s(v_x, v_y)$ of the SG2d-model with only active fluctuations for different noise strengths: $D_v = 0.1$, top; $D_v = 0.2$, center; $D_v = 0.5$ bottom. **Left column:** Results obtained from Langevin simulations; **Central column:** Analytical solution given in Eq. 3.28; **Right column:** One dimensional cross-sections $P_s(v_x, 0)$ comparing analytical solutions (solid lines) with numerics (symbols); Other parameters $\alpha = 1.0, v_0 = 1.0$.

with $_1F_1(a, b, z)$ being the Kummer's confluent hypergeometric function, $U(a, b, z)$ being the simple confluent hypergeometric function, $\Gamma(z)$ being the Euler gamma function, and

$$A = \frac{D_v}{2(D + D_v)}, \quad B = \frac{D}{2(D + D_v)}, \quad C = \frac{\alpha v_0^2}{2(D + D_v)} \, . \tag{3.32}$$

The speed PDF can be derived in analogy to the purely passive/active fluctuation case to:

$$\tilde{p}_m(|v|) = \mathcal{N}_m |v|^{\frac{D}{D+D_v}} e^{-\frac{\alpha(|v|-v_0)^2}{2(D+D_v)}} \left[1 + e^{-\frac{2\alpha|v|v_0}{(D+D_v)}} \right] \, . \tag{3.33}$$

It increases for low speeds as $|v|^c$ with $c = D/(D + D_v)$ ($0 \leq c \leq 1$). The extreme c-values correspond to the limiting cases of purely active fluctuations $c = 0$ and purely passive fluctuations $c = 1$. The probability PDF assumes a finite value at vanishing speed ($\tilde{p}(0) > 0$) only for purely active fluctuations $D = 0$, but we observe for $D > 0$ an increasing probability[1] of low speeds with increasing strength of active velocity fluctuations D_v (Fig. 3.3).

Our results show that already for the simple Stokes friction active velocity fluctuations lead to characteristic deviations from the classical Rayleigh PDF (Eq. 3.20). Active velocity fluctuations acting along the direction of motion increase the probability of low speeds $|v| \ll 1$. Although, we have only derived analytical results for the Schienbein-Gruler friction, which reduces to Stokes friction in the limit $v_0 = 0$, the qualitative result is of general nature and holds for arbitrary friction functions. The initial nonlinear increase of the speed distribution leads to a divergence of the corresponding Cartesian velocity distributions. Even for the linear Stokes friction the Cartesian velocity distributions are not simple Gaussian but show a distinct peak at the origin ($v_x, v_y = 0$) similar to the one shown in the bottom panels of Fig. 3.5. Such distributions have in fact been reported from experiments on cell motility by Bödeker et al. (2010). We expect that a detailed analysis of experimental data for different active systems, such as chemically powered colloids, will yield similar results. In fact the exponent $c < 1$ may be used as an indicator for the presence of active fluctuations in systems far from equilibrium.

---

[1] integral of speed PDF over a finite interval

3.2. Spatial Diffusion

3.2.1. Only Passive Fluctuations

The mean square displacement (MSD) of an active particle $\Delta\mathbf{r}^2(t) = (\mathbf{r}(t) - \mathbf{r}(0))^2$ is given by:

$$\Delta\mathbf{r}^2(t) = 2 \int_0^t dt' \int_0^{t'} d\tau \langle \dot{\mathbf{r}}(0)\dot{\mathbf{r}}(\tau)\rangle . \tag{3.34}$$

According to the equations of motion the correlation function can be rewritten as:

$$\begin{aligned}\langle \dot{\mathbf{r}}(0)\dot{\mathbf{r}}(\tau)\rangle &= \langle v(0)v(\tau)\mathbf{e}_h(0)\mathbf{e}_h(\tau)\rangle \\ &= \langle v(0)v(\tau)\cos(\varphi(\tau) - \varphi(0))\rangle . \end{aligned} \tag{3.35}$$

In contrast to the one-dimensional case, the heading dynamics in two dimensions are not independent from the velocity dynamics, as the effective coefficient of angular diffusion depends on v.

Assuming constant velocity $v = const$ the Eq. (3.15) reduces to

$$\frac{\partial p(\varphi, t)}{\partial t} = \frac{D}{v^2}\frac{\partial^2 p(\varphi, t)}{\partial\varphi^2}. \tag{3.36}$$

With the initial condition $p(\varphi, t_0) = \delta(\varphi - \varphi(t_0))$ and periodic boundary taking into account the periodicity of the heading $p(-\pi, t) = p(\pi, t)$ the time dependent solution of (3.36) reads

$$p(\varphi(\tau), \tau|\varphi(0), 0) = \frac{1}{\pi}\left(\frac{1}{2} + \sum_{n=1}^{\infty}\cos\left[n(\varphi - \varphi(t_0))\right]e^{-n\frac{D}{v^2}\tau}\right) . \tag{3.37}$$

For large times $(t \to \infty)$ the distribution $p(\varphi, t)$ converges towards the homogeneous distribution $1/(2\pi)$, which corresponds to a complete loss of directional information. The relaxation rate of the angular function in (3.36) is given by the relaxation rate of the first Fourier mode $(n = 1)$.

For constant velocity $v = v_0$ we may rewrite the correlation function as

$$\langle v(0)v(\tau)\cos(\varphi(\tau) - \varphi(0))\rangle = v_0^2 e^{-\frac{D}{v^2}\tau}. \tag{3.38}$$

Inserting Eq. (3.38) in Eq. (3.34) yields

$$\Delta\mathbf{r}^2(t) = 2\frac{v_0^4}{D}\left[t + \frac{v_0^2}{D}\left(e^{-\frac{D}{v_0^2}t} - 1\right)\right] . \tag{3.39}$$

This is a well known result obtained previously by Mikhailov and Meinköhn (1997). The corresponding effective diffusion coefficient is inversely proportional to D and reads

$$D_{\mathrm{MM}} = \lim_{t\to\infty}\frac{\Delta\mathbf{r}^2(t)}{4t} = \frac{v_0^4}{2D} . \tag{3.40}$$

In the limit $D \to \infty$ the effective diffusion coefficient vanishes, i.e. $D_{\mathrm{MM}} \to 0$.

For the SG2d-model at low noise intensities $D/\alpha \ll v_0^2$ the velocity distribution is given by a narrow peak around v_0. As a first approximation the velocity can be assumed to be constant and the external fluctuation act only on the direction of motion. Thus in the limit $D \to 0$ the effective diffusion coefficient will converge to Eq. 3.40. In the limit of large noise intensities $D/\alpha \gg v_0^2$ we may neglect the active motion term in the SG-friction $v_0 \to 0$. The dynamics reduce to ordinary Brownian motion with the effective diffusion coefficient $D_{\mathrm{BM}} = D/\alpha^2$ which increases linearly with D and vanishes for $D \to 0$.

Based on the two asymptotic limits it becomes evident that there must exist a minimum of the effective diffusion coefficient. A simple approximation for $D_{\mathrm{eff}}^{(2d)}$ can be obtained by a sum of the two asymptotic diffusion coefficients:

$$D_{\mathrm{I}} = D_{\mathrm{MM}} + D_{\mathrm{BM}} = \frac{v_0^4}{2D} + \frac{D}{\alpha^2} \ . \tag{3.41}$$

This approximation has the right asymptotic behavior and reproduces qualitatively the behavior of $D_{\mathrm{eff}}^{(2d)}$ at intermediate noise strengths D. However, a comparison with the numerical results reveals that this approximation underestimates the diffusion coefficient close to the minimum (Fig. 3.6).

An alternative approximation of $D_{\mathrm{eff}}^{(2d)}$ can be obtained by considering the velocity drift term in the Fokker-Planck Equation (3.15) with $v_0 > 0$. The most probable velocities $\tilde{v}$, which correspond to the maxima of the $p_s(v)$, are given as roots of the drift term including the Stratonovich shift:

$$-\alpha(\tilde{v} - v_0) + \frac{D}{\tilde{v}} = 0 \tag{3.42}$$

By multiplying (3.42) with $\tilde{v}$ we obtain a quadratic equation for $\tilde{v}$ with the roots:

$$\tilde{v}_{+/-} = \frac{v_0}{2} \pm \sqrt{\frac{v_0^2}{4} + \frac{D}{\alpha}} \ . \tag{3.43}$$

The positive root corresponds to the maximum at positive velocities close to v_0, whereas the negative root corresponds to the maximum at negative velocities (backward motion). For small D the backward motion may be neglected and the most probable velocity is given by the positive root $\tilde{v}_+$. Inserting $v = \tilde{v}_+$ in the expression for D_{MM} (3.40) yields the right qualitative behavior of the diffusion coefficient with a minimum at intermediate D, but lacks the correct asymptotic behavior for $D \to \infty$. In order to eliminate this deviation we add a correction term $D/(2\alpha^2)$ and obtain a second approximation:

$$D_{\mathrm{II}} = \frac{\left(\frac{v_0}{2} + \sqrt{\frac{v_0}{4} + \frac{D}{\alpha}} \right)^4}{2D} + \frac{D}{2\alpha^2} \ . \tag{3.44}$$

The comparison of D_{II} with numerical results shows that it offers a better approximation then the D_{I} but overestimates the diffusion coefficient close to the minimum.

The two approximations provide a lower and an upper bound of the effective diffusion coefficient close to the minimum. Thus, by taking the average of D_{I} and D_{II} we obtain a third approximation for $D_{\text{eff}}^{(2d)}$. This heuristic ansatz does not yield any additional qualitative insights but results in an analytical expression for $D_{\text{eff}}^{(2d)}$ with a good agreement to numerical simulation:

$$D_{\text{eff}}^{(2d)} \approx D_{\text{III}} = \frac{1}{2} \left(D_{\text{I}} + D_{\text{II}} \right)$$

$$= \frac{3D}{4\alpha^2} + \frac{v_0^4}{4D} + \frac{1}{64D} \left(v_0 + \sqrt{v_0^4 + \frac{4D}{\alpha}} \right)^4 . \tag{3.45}$$

In Fig. 3.6 we show a comparison of the three analytical approximations for $D_{\text{eff}}^{(2d)}$. Please note that at low noise intensities all approximations yield systematically larger values of $D_{\text{eff}}^{(2d)}$ than obtained from numerical simulations. This can be explained by the coupling of the effective angular diffusion to the velocity dynamics. In the next section we will consider the Schienbein-Gruler model with active fluctuations. Although the $v\varphi$-coupling is also present for active fluctuations it is possible to consider limiting cases where v and φ-dynamics effectively decouple.

3.2.2. Only Active Fluctuations

For small velocity fluctuations D_v the mean squared displacement $\langle \Delta \mathbf{r}^2 \rangle$ can be calculated in analogy to the external noise case by assuming constant velocity $v(t) = v_0 = const.$ The effective diffusion coefficient is given by the Mikhailov and Meinköhn (1997) formula (3.40): $D_{\text{eff}}^{(2d)} = D_{\text{MM}} = v_0^4/(2D_\varphi)$.

In general, the angular diffusion depends explicitly on the velocity D_φ/v^2 and for larger velocity fluctuations the approximation of independent velocity and angular dynamics fails. Furthermore, in contrast to the external noise case, the dynamics do not converge towards ordinary Brownian motion in the limit of large D_v, D_φ. Thus, we chose a different approach in order to obtain approximate expressions for the coefficient of effective diffusion.

Recently, Peruani and Morelli (2007) showed that, assuming independent angular and velocity dynamics with exponential decay of correlations, the mean squared displacement of self-propelled particles is given by:

$$\langle \Delta \mathbf{r}^2 \rangle(t) = 4D_{\text{eff}}t = 2\frac{\langle v \rangle^2}{\kappa_\varphi^2} \left(\kappa_\varphi t - 1 + e^{-\kappa_\varphi t} \right)$$

$$+ 2\frac{\langle v^2 \rangle - \langle v \rangle^2}{(\kappa_\varphi + \kappa_v)^2} \left((\kappa_\varphi + \kappa_v)t - 1 + e^{-(\kappa_\varphi + \kappa_v)t} \right) . \tag{3.46}$$

Here, $\kappa_{v,\varphi}$ are the relaxation rates of velocity and angle given by the inverse correlation times, whereas $\langle v \rangle$ and $\langle v^2 \rangle$ are the first two moments of the stationary velocity distribution. The corresponding effective diffusion coefficient reads

$$D_{\text{eff}}^{(2d)} = \lim_{t \to \infty} \frac{\langle \Delta \mathbf{r}^2 \rangle(t)}{4t} = \frac{1}{2\kappa_\varphi} \left[\langle v \rangle^2 + \frac{\langle v^2 \rangle - \langle v \rangle^2}{(1 + \frac{\kappa_v}{\kappa_\varphi})} \right] . \tag{3.47}$$

The above result corresponds to the one obtained for the turning model in one dimension 2.21. For the SG model with active fluctuations we may immediately identify three of the

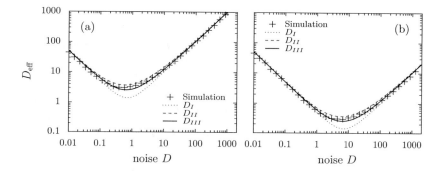

Figure 3.6.: Effective diffusion coefficient of SG2d-model with only passive fluctuations versus noise intensity D for two different parameter sets: (a) $\alpha = 1.0$ and $v_0 = 1.0$, (b) $\alpha = 10.0$ and $v_0 = 1.0$. Comparison of numerical results with the analytical approximations D_I (3.41), D_{II} (3.44) and D_{III} (3.45).

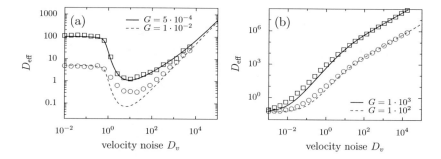

Figure 3.7.: Effective diffusion coefficient of SG2d-model with active fluctuations versus noise intensity D_v for different parameter values. (a) Fast velocity dynamics and slow angular dynamics: $\alpha = 10.0$, $v_0 = 1.0$, $v_\varepsilon = 0.01$ and $D_\varphi = 5 \cdot 10^{-3}$ (black), $1 \cdot 10^{-2}$ (red); (b) Slow velocity dynamics and fast angular dynamics: $D_\varphi = 10.0$, $v_0 = 1.0$, $v_\varepsilon = 0.001$ and $\alpha = 0.01$ (black), 0.01 (red). The solid/dashed lines represent the theoretical predictions obtained using Eqs. 3.47 and (a) 3.52, (b) 3.54

four parameters in Eq. 3.47:

$$\langle v \rangle = v_0, \qquad \langle v^2 \rangle = v_0^2 + \frac{D_v}{\alpha}, \qquad \kappa_v = \alpha, \tag{3.48}$$

whereas the angular relaxation rate is less obvious. Actually, due to the coupling of the angular diffusion to the velocity the underlying assumption used in the derivation of Eqs. 3.46, 3.47 is violated. Despite this fundamental discrepancy some limiting cases allow us to consider the angular and velocity dynamics as effectively decoupled and enables the derivation of approximation for $D_{\text{eff}}^{(2d)}$ using Eq. 3.47. The first example, already discussed above, is the case of vanishing velocity fluctuations ($\alpha \to \infty$). Another example is a large time scale separation of angular and velocity dynamics for $v \approx v_0$: $\alpha \gg D_\varphi / v_0^2$. In this case, it may be assumed that for a slowly changing $\varphi(t) \approx const.$ the fast velocity relaxation ensures the stationarity of the velocity distribution.

In the point-like particle model this assumption is, strictly speaking, always violated due to the divergence of the angular dynamics for $v \to 0$. This leads to an extremely fast angular diffusion if, due to fluctuations, the velocity comes close to 0. Here, we eliminate this unphysical divergence introducing a limiting angular diffusion at small velocities $|v| < v_\varepsilon \ll 1$, where $v_\varepsilon > 0$ is a constant defining the cut-off velocity in the angular dynamics. The modified angular equation of motion reads

$$\dot{\varphi} = H_c(v)\sqrt{2D_\varphi}\xi_\varphi \,, \tag{3.49}$$

with

$$H_c(v) = \begin{cases} \frac{v}{|v|v_\varepsilon} & \text{for} \quad |v| < v_\varepsilon \\ \frac{1}{v} & \text{for} \quad |v| > v_\varepsilon \end{cases} . \tag{3.50}$$

The above definition of H_c ensures that the sign of $\dot{\varphi}$ – thus the turning direction due to applied force – is always the same as in the original point-like particle model ($H_c \lessgtr 0$ if $v \lessgtr 0$). For $|v| > v_\varepsilon$ the modified angular equation is identical to the point-like case. The angular diffusion remains finite for arbitrary velocities. It is constant below the cut-off velocity v_ε due to the constant absolute value of $H_c(v)$:

$$|H_c(v)| = \frac{1}{v_\varepsilon} = const. \qquad \text{for} \qquad |v| < v_\varepsilon \,. \tag{3.51}$$

The separation of time scales can be expressed by the dimensionless number $G = D_\varphi/(\alpha v_0^2)$. For $G \ll 1$ and $v \approx v_0$ the velocity relaxation is much faster than the decorrelation of φ, whereas for $G \gg 1$ the velocity dynamics is slow compared to the diffusive time scale of the angular degree of freedom.

For the modified model, under the assumption of fast velocity relaxation ($G \ll 1$), we make the following ansatz for κ_φ:

$$\kappa_\varphi = \langle H_c^2 D_\varphi \rangle = \int_0^\infty d|v| \; H_c^2 D_\varphi \tilde{p}_a(|v|)$$
$$= \int_0^\infty d|v| \; H_c^2 D_\varphi \mathcal{N}_a e^{-\frac{\alpha(|v|-v_0)^2}{2D_v}} \left(1 + e^{-\frac{2v_0|v|}{2D_v}}\right) \,. \tag{3.52}$$

It is not possible to obtain an analytical solution of the above integral but it can be easily solved with standard numerical integration schemes, due to sufficiently fast convergence with increasing upper integration limit.

Please note that κ_φ depends indirectly on the velocity noise D_v via its dependence on $\tilde{p}_a(|v|)$: With increasing D_v the probability of lower speeds – at which fast angular diffusion takes place – increases initially. On the other hand, with increasing D_v, the total width of $p_a(v)$ increases as well. Thus, at some point a further increase in D_v results in the decrease of the relative frequency of low speeds. As a consequence κ_φ starts to decrease. The resulting maximum of κ_φ versus D_v corresponds to a minimum of the spatial diffusion coefficient $D_{\text{eff}}^{(2d)}$ (3.47) (Fig. 3.7). The strong inhibiting impact of velocity fluctuations on the mean squared displacement of individual trajectories is shown in Fig. 3.8.

Comparing the theoretical results (Eqs. 3.47 and 3.52) to numerical simulations of the modified model yields good qualitative and quantitative agreement for large time scale separations (Fig. 3.7a). With decreasing separation of time scales the numerical results show similar qualitative behavior but the minimum of $D_{\text{eff}}^{(2d)}$ is less pronounced than predicted by the theory. For large $D_\varphi \sim G$ the minimum of $D_{\text{eff}}^{(2d)}$ disappears.

For fast angular dynamics (and $v_\varepsilon \gtrsim 0$) we make the following ansatz for κ_φ:

$$\kappa_\varphi = \langle \frac{D_\varphi}{v^2} \rangle = \frac{D_\varphi}{\langle v^2 \rangle} = \frac{D_\varphi}{v_0^2 + \frac{D_v}{\alpha}} \ . \tag{3.53}$$

Inserting Eq. 3.53 in Eq. 3.47 yields

$$D_{\text{eff}}^{(2d)} = \frac{1}{2} \left(\frac{v_0^2}{D_\varphi} + \frac{D_v}{\alpha(D_\varphi + D_v + \alpha v_0^2)} \right) \left(v_0^2 + \frac{D_v}{\alpha} \right) \ . \tag{3.54}$$

This approximation is, strictly speaking, only valid for the point-particle model $v_\varepsilon = 0$ but it is also in good quantitative and qualitative agreement with the numerical results of the modified model for small v_ε (see Fig. 3.7b).

Finally, by setting $D_v = D_\varphi = D_a$, we may simplify Eq. 3.54 to

$$D_{\text{eff,a}}^{(2d)} = \frac{(D_a + \alpha v_0^2)^3}{2\alpha^2 D_a (2D_a + \alpha v_0^2)} \ . \tag{3.55}$$

Setting angular and velocity noise strengths equal allows for $v_\varepsilon \to 0$ a direct comparison of the effective diffusion coefficients for active and passive fluctuations for the same total noise strengths $D = D_a$. The diffusion coefficient obtained from Eq. (3.55) exhibits a minimum with respect to D_a as in the passive fluctuation case, but for $D_a \to \infty$ it does not converge towards the result expected for ordinary Brownian diffusion $D_a/(\alpha^2)$ but to a lower value $D_a/(4\alpha^2)$. Thus for comparable total noise strength the diffusive dispersal of point particles subject to active fluctuations is always slower then the dispersal due to only passive fluctuations. The direct comparison of $D_{\text{eff}}^{(2d)}$ for active and passive fluctuations confirms our findings (Fig. 3.9). The result obtained from (3.55) is in a good agreement with the numerical simulation results, but shows a systematic deviation at moderate noise strengths. This is due to the coupling of the angular degree of freedom to the velocity, which is not taken properly into account in Eq. 3.55.

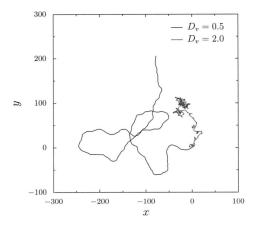

Figure 3.8.: Examples of single particle trajectories for the SG-model with active fluctuations for two different values of the velocity noise $D_v = 0.5,\ 2.0$. Both trajectories start at $t = 0$ at $(x, y) = (0, 0)$ with random initial direction of motion and $v(t = 0) = v_0$, the total simulation time is $t = 1000$. Other parameters values: $\alpha = 10.0,\ v_0 = 1.0,\ D_\varphi = 0.05$.

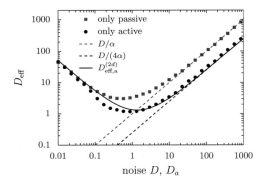

Figure 3.9.: Comparison of the effective diffusion coefficient of the SG2d-model for active fluctuations (black) and passive (red) fluctuations versus the corresponding noise strengths $D,\ D_a$. The symbols show the results of Langevin simulations. The solid line shows the results obtained from Eq. (3.55), whereas the dashed lines indicate the asymptotic scaling for $D, D_a \to \infty$. Other parameter values: $\alpha = 1.0,\ v_0 = 1.0$.

4. Speed Model

In this chapter we consider active Brownian particles, or agents, in two dimensions with speed $s = |\mathbf{v}|$ and direction of motion (heading) as degrees of freedom in contrast to the velocity model in Chapter 3. Although this distinction might seem gratuitous, it has some important consequences. Consideration of the speed $s \geq 0$ and not the velocity $v = \mathbf{v}\mathbf{e}_h$, which may be negative, as a degree of freedom simplifies the mathematical treatment and allows to define a general form of a propulsion function (negative friction function).

A speed-model is also a reasonable ansatz for the description of biological motion where most (higher) organisms not only have a clearly defined "head-tail" asymmetry but also move (almost) exclusively in the direction of the "head". Most biological agents prefer to turn around when responding to external cues and move in the direction of their "head" instead of performing a backward motion.

We restrict the discussion to spatially and temporarily homogeneous dynamics. In the speed model the motion of an active agent in arbitrary spatial dimension d is described by

$$\dot{\mathbf{r}} = \mathbf{v} = s(t)\mathbf{e}_h(t). \tag{4.1}$$

Here, $\mathbf{r}(t)$ and $\mathbf{v}(t)$ are the position and the velocity vector of the agent at time t ($\mathbf{r}, \mathbf{v} \in \mathbb{R}^d$), $s(t)$ is the time dependent speed of the agent and $\mathbf{e}_h(t)$ is the unit vector in the direction of motion. The details of the temporal evolution of the direction of motion depend strongly on the dimension of the problem. First we will focus on the speed dynamics which are independent of the dimension of the motion.

We make the following ansatz for the temporal evolution of the speed

$$\dot{s} = f(s) + \eta(t) . \tag{4.2}$$

The first term on the right-hand side is an arbitrary speed-dependent propulsion function $f(s)$, which describes the deterministic evolution of the speed. The second term is the stochastic part given by a random force[1] $\eta(t)$. In general, the random force $\eta(t)$ can have arbitrary distributions and temporal statistics. For simplicity we assume the speed noise to be white Gaussian with intensity D_s: $\eta(t) = \sqrt{2D_s}\xi_s$ (2.5), which can be considered based on the center limit theorem as a reasonable approximation of many random processes occuring in nature (Gardiner, 1985).

We have to ensure that $s(t)$ is positive at all times. This can be achieved, for example, through an appriopriate choice of $f(s)$ with diverging acceleration for $s \rightarrow 0$. Here we chose a rather simple approach by imposing a reflecting boundary conditions at $s = 0$.

Please note that the speed model with reflective boundary condition is equivalent to the

---

[1] Please note that we use here the term "force" not in a strict physical sense. With "force", we refer to effective deterministic and stochastic terms, which lead to an acceleration or deceleration of an active particle.

velocity model if the propulsion function $f(v)$ is symmetric with respect to $v = 0$, such as the negative Rayleigh-Helmholtz friction (Erdmann et al., 2000). For such symmetric propulsion functions with v as a degree of freedom, velocity noise may induce transitions over $v = 0$, which correspond to abrupt changes in direction. Thus, the intensity of the velocity fluctuations would have a strong impact on the direction of motion. The decoupling of the speed dynamics from the direction of motion is another argument for considering speed – and not velocity – as a degree of freedom.

The mean squared displacement (MSD), and thus the effective diffusion coefficient D_{eff}, of an active Brownian particle in arbitrary dimensions depends strongly on its speed dynamics. For a stationary process the MSD is determined by the moments of the stationary speed distribution $p(s)ds$ (Peruani and Morelli, 2007). In general, the stationary probability density function (PDF) reads

$$p(s) = \mathcal{N}e^{-V(s)/D_s} \qquad \text{with} \qquad \mathcal{N}^{-1} = \int_0^\infty ds' \exp\left(-\frac{V(s')}{D_s}\right) . \qquad (4.3)$$

$\mathcal{N}$ is the normalization constant and $V(s) = -\int_0^s ds' f(s')$ is the effective speed potential. The existence of a finite $\mathcal{N}$ (normalization condition) defines the general restriction for the choice of $f(s)$.

The expectation value $\langle s^n \rangle$ (n-th moment of the speed distribution) is formally given as

$$\langle s^n \rangle = \int_0^\infty ds' \, s'^n p_s(s'). \qquad (4.4)$$

We define a propulsion function $f(s)$ of active motion as any function that has at least one stable fixed point for $s > 0$ and $D_s \to 0$ and ensures that the speed dynamics do not diverge at large speeds: $\lim_{s \to \infty} f(s) < 0$. Here we will restrict to simple propulsion functions with only one stable fixed point $s_0 > 0$, with $f(s_0) = 0$ and $f'(s_0) < 0$. Thus for vanishing noise, the speed of the agent will always relax towards s_0.

The behavior of the first two speed moments $\langle s^n \rangle$ ($n = 1, 2$) with respect to the noise intensity D_s can be analyzed without explicitly solving Eq. 4.4, which might be rather complicated or even impossible for an arbitrary $f(s)$. The only assumption is that $f(s)$ is continuous and differentiable at s_0.

First, for small D_s the speed of an agent in Eq. 4.2 is given by the stationary speed for $D_s = 0$ plus some small deviation: $s = s_0 + \delta s$. Thus the first two moments can be rewritten as:

$$\langle s \rangle = s_0 + \langle \delta s \rangle \qquad (4.5a)$$

$$\langle s^2 \rangle = s_0^2 + 2\langle \delta s \rangle s_0 + \langle \delta s^2 \rangle \qquad (4.5b)$$

Now we expand $f(s)$ up to second order around s_0 and subsitute our assumption for s in Eq. 4.2, which yields

$$\frac{d}{dt}\delta s = g_1 \delta s + \frac{g_2}{2}\delta s^2 + \xi_v(t), \qquad (4.6)$$

where we used the abbrivations $g_1 = f'(s_0)$ and $g_2 = f''(s_0)$.

The Fokker-Planck equation for the distribution of speed deviations $Q(\delta s, t)$ reads

$$\frac{\partial Q(\delta s, t)}{\partial t} = -\frac{\partial}{\partial s}\left(g_1 \delta s Q + \frac{g_2}{2}\delta s^2 Q - D_s \frac{\partial}{\partial s}Q\right). \tag{4.7}$$

Inserting the Fokker-Planck equation in Eq. 4.4 gives us, after some calculus, the differential equations for the temporal evolution of the speed moments:

$$n = 1 \quad \frac{d}{dt}\langle \delta s \rangle = g_1 \langle \delta v \rangle + \frac{g_2}{2}\langle \delta s^2 \rangle, \tag{4.8a}$$

$$n = 2 \quad \frac{d}{dt}\langle \delta s^2 \rangle = g_1 \langle \delta s^2 \rangle + \frac{g_2}{2}\langle \delta s^3 \rangle + 2D_s, \tag{4.8b}$$

$$n \geq 2 \quad \frac{d}{dt}\langle \delta s^n \rangle = g_1 \langle \delta s^n \rangle + \frac{g_2}{2}\langle \delta s^{n+1} \rangle + n(n-1)D_s \langle s^{n-2} \rangle. \tag{4.8c}$$

As can be seen from the above expressions, the evolution of the n-th moment, and therefore also its stationary value, depends on the (n+1)-th moment. Thus, without any assumptions on the distribution Q, the moment equations form an infinite hierarchy. Here we impose a closure by setting $\langle \delta s^3 \rangle = 0$, which is a reasonable approximations at small noise intensities. With this approximation from Eqs. 4.8 the first two stationary moments of the speed deviations read:

$$\langle \delta s \rangle = \frac{D_s g_2}{g_1^2}, \tag{4.9a}$$

$$\langle \delta s^2 \rangle = -\frac{2D_s}{g_1} > 0 . \tag{4.9b}$$

Finally, inserting the above results into the expressions for the speed moments (4.5) yields

$$\langle s \rangle = s_0 + \frac{g_2}{g_1^2}D_s, \tag{4.10}$$

$$\langle s^2 \rangle = s_0^2 + \frac{2D_s}{g_1^2}(s_0 g_2 - g_1). \tag{4.11}$$

The behavior of $\langle s \rangle$ with increasing (but small) D_s is determined by the sign of the second derivative: For $g_2 > 0$ ($f(s)$ is convex) the mean speed increases with increasing D_s. For $g_2 < 0$ ($f(s)$ is concave) the mean speed decreases with increasing noise. The impact of noise on the second moment depends on the sign of $A = s_0 g_2 - g_1$: For $A > 0$ ($A < 0$) the second moment increases (decreases) with increasing D_s.

It should be emphasized that the above considerations do not account for the reflecting boundary and hold only in the limit of small D_s. The reflecting boundary at $s = 0$ puts a lower bound on $\langle \delta s \rangle$. In the limit of large D_s the probability of speeds larger than s_0 in the absence of any upper bound on s increases continuously with D_s. In parallel the probability of speeds in the interval $0 < s < s_0$ has to decrease. Therefore at large D_s the mean speed $\langle s \rangle$ is an increasing function of D_s. Thus, we conclude that for a concave propulsion functions $g_1 < 0$ a minimum of $\langle s \rangle$ as a function of D_s exists. The same applies to the second moment and for $A < 0$ a minimum of $\langle s^2 \rangle$ exists, whereas for $A \geq 0$ the second moment increases monotonously with D_s.

4.1. Propulsion Functions of Active Motion

In modelling of active Brownian motion different speed dependent propulsion functions were proposed. The most prominent examples are the Schienbein-Gruler function (SG) (Schienbein and Gruler, 1993) derived from cell experiments, the Schweitzer-Ebeling-Tilch function (SET) (Schweitzer et al., 1998; Erdmann et al., 2000) derived from theoretical considerations of particles with an internal energy depot, as well as, the Rayleigh-Helmholtz friction (RH) (Rayleigh, 1894; Erdmann et al., 2000). The different propulsion functions have in common that by changing of a single (bifurcation) parameter[2] a stable fixed point of the deterministic dynamics at $s = 0$ becomes unstable and a new stable fixed point emerges at finite speed $s_0 > 0$ with $f(s_0) = 0$ and $f'(s_0) < 0$. In this case, in the absence of noise, any initial condition $s(0) > 0$ will relax towards s_0 and we may refer to this situation as the active motion regime. By rescaling the speed s by the corresponding stationary speed s_0 for the different friction functions, we may rewrite those functions in the active motion regime to (see Appendix A):

$$f_{SG}(s) = \alpha(1 - s), \tag{4.12}$$

$$f_{RH}(s) = \alpha(s - s^3), \tag{4.13}$$

$$f_{SET}(s) = \alpha \left(\frac{\beta s}{s^2(\beta - 1) + 1} - s \right). \tag{4.14}$$

Due to the rescaling the speed s becomes dimensionless. The SG and RH functions depend only on a single parameter α, with the unit of inverse time, whereas in the SET-model a second dimensionless parameter, $\beta > 1$, remains. For $\beta \gg 1$ and $s \gg 0$ this additional parameter can be eliminated and the SET-model can be approximated as

$$f_{aSET}(s) = \alpha \left(\frac{1}{s} - s \right) \tag{4.15}$$

Please note that the rescaling of the speed in Eq. 4.2 also rescales the noise strength according to $D_s \to D_s/s_0^2$. Examples of the corresponding speed distributions determined by Eq. 4.3 are shown in Fig. 4.1. Through a comparison of the structure of the different friction functions, SG (4.12), RH (4.13) and aSET(4.15)), as well as the requirement of the existence of a fixed point at $s = 1$ ($g_1 < 0$), a general propulsion function can be introduced:

$$f_{(gen)}(s) = \alpha(s^a - s^b) = \alpha s^b(s^{a-b} - 1), \tag{4.16}$$

with $a < b$.

The values of g_2 and A from the previous section determine the behavior of the first

---

[2]In contrast to the RH and SET propulsion functions, there is no qualitative change of the phase space in the SG-Model, thus no bifurcation takes place. A change in a corresponding parameter in the SG-model shifts the only stable fixed point from $s = 0$ to $s \neq 0$.

and second moment. For the different propulsion functions they read

$$g_{2,SG} \quad = 0 \qquad\qquad A_{SG} \quad = \alpha > 0, \qquad (4.17)$$

$$g_{2,RH} \quad = -6\alpha < 0 \qquad\qquad A_{RH} \quad = -4\alpha < 0, \qquad (4.18)$$

$$g_{2,aSET} = 2\alpha > 0 \qquad\qquad A_{sSET} = 4\alpha > 0, \qquad (4.19)$$

$$g_{2,SET} \quad = \frac{2\alpha(\beta - 4)(\beta - 1)}{\beta^2} \qquad A_{SET} \quad = \frac{4\alpha(\beta - 2)(\beta - 1))}{\beta^2}. \qquad (4.20)$$

For the general single parameter propulsion function we obtain:

$$g_{2,\text{gen}} = \alpha(a^2 - b^2 - (a - b)) \qquad A_{\text{gen}} = \alpha(a^2 - b^2 - 2(a - b)). \qquad (4.21)$$

By applying the result from the previous section we predict the existence of a minimum of the speed moments $\langle s^n \rangle$ for the RH-model in contrast to monotonous increase of the first two moments for the SG-model and the aSET-model. Please note that for the SG-model we obtain $g_{2,SG} = 0$ which, in the absence of a reflecting boundary at $s = 0$ (SG velocity model, see chapter 3), would yield a first moment independent on D_s, but the reflecting boundary results in increasing $\langle s \rangle$ for sufficiently large D_s.

For the general propulsion function, depending on the exponents a and b, we can identify three different situations (see Fig. 4.2):

A) Both moments ($\langle s \rangle$, $\langle s^2 \rangle$) have a minimum with respect to D_s;

B) $\langle s \rangle$ has a minimum but $\langle s^2 \rangle$ has no minimum with repect to D_s;

C) Both moments increase monotonously with D_s.

The SET function may fall into either of the three regimes depending on the value of the parameter β: A) for $1 < \beta < 2$, B) for $2 < \beta < 4$ and C) for $\beta > 4$.

For the RH and SG models as well as for the approximated SET-model (aSET) the moments of the stationary speed distribution can be obtained as analytical expressions (see Appendix B). The directly calculated moments are in agreement with our prediction on their behavior with respect to increasing noise intensity D_s. In addition to the propulsion functions, which have already been discussed in the literature, we analyzed the moments of a function with the exponents $a = 1/2$ and $b = 1$ which we will refer to as SL-friction function (*square-root+linear* propulsion function). Here we omit the analytic result for either of the first two moments obtained from solution of Eq. 4.4 for the SL-friction as they are given by complex expressions involving hypergeometric series. For the SL-friction our results predict a minimum of the first moment and a monotonously increasing second moment with increasing D_s, which is confirmed by the direct result.

4.2. Diffusion of Active Brownian Agents in 2D

We consider the motion of an active Brownian agent in two spatial dimensions with $\dot{\mathbf{r}} = \mathbf{v}$. The velocity vector of an agent at time t is given as

$$\mathbf{v} = s(t)\mathbf{e}_h(t) = (s(t)\cos\varphi(t), s(t)\sin\varphi(t)) \qquad (4.22)$$

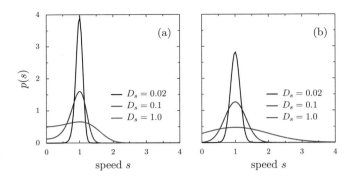

Figure 4.1.: Examples of the speed distributions for a) RH friction and b) SG friction for different values of the speed noise $D_s = 0.02$, 0.1 and 1.0 ($\alpha = 1.0$).

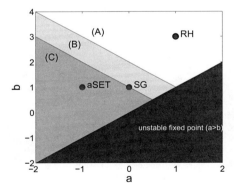

Figure 4.2.: Behavior of the first two moments in dependence of the exponents a and b of the generalized propulsion function of active motion (4.16). Three situation can be distinguished: (A) Both moments ($\langle s \rangle$, $\langle s^2 \rangle$) have a minimum vs D_s; (B) $\langle s \rangle$ has a minimum but $\langle s^2 \rangle$ has no minimum vs D; (C) Both moments increase monotonously with D_s. The points represent the position of RH, $aSET$ and SG propulsion functions.

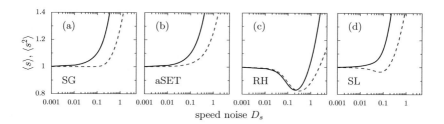

Figure 4.3.: The first moment $\langle s \rangle$ (solid line) and the second moment $\langle s^2 \rangle$ (dashed line) of the speed in dependence on noise intensity D_s for different friction functions ($\alpha = 1$): a) Schienbein-Gruler (SG), b) approximated Schweitzer-Ebeling-Tilch (aSET), c) Rayleigh-Helmholtz (RH) and the square-root+linear (SL) friction function.

with $\varphi(t)$ determining the heading of the agent with respect to a fixed reference angle.

The speed dynamics are given by Eq. 4.2 and we need now to define the dynamics of the angle $\varphi(t)$. In the following, we will discuss two different examples of angular dynamics: *run & tumble* and continuous angular diffusion due to (internal) angular fluctuations (see also Ch. 3).

4.3. Run and Tumble

We consider first a *run & tumble* motion of individual agents: The agent moves with a fluctuating speed determined by a propulsion function Eq. 4.2 in a certain direction φ (*run*). With a rate κ_φ discrete reorientation events (*tumbles*) take place where a new direction of motion is randomly chosen from a uniform distribution $\varphi \in [0, 2\pi]$. Thus, each particle performs successive straight runs with with varying speed, uncorrelated directions and exponentially distributed durations.

The speed and orientation dynamics are completely independent and the temporal correlations in the direction of motion are exponentially decaying. With the additional assumption of exponentially decaying speed correlations we may directly apply the results for the mean squared derived by Peruani and Morelli (2007) Eq. (3.46). The spatial diffusion coefficient reads (in analogy to Eq. 3.47):

$$D_{\text{eff}}^{(2d)} = \lim_{t \to \infty} \frac{\langle \Delta \mathbf{r}^2 \rangle(t)}{4t} = \frac{1}{2\kappa_\varphi} \left[\langle s \rangle^2 + \frac{\langle s^2 \rangle - \langle s \rangle^2}{(1 + \frac{\kappa_s}{\kappa_\varphi})} \right]. \tag{4.23}$$

Here, $\kappa_{s,\varphi}$ are the relaxation rates of the speed and the angle given by the corresponding inverse correlation times.

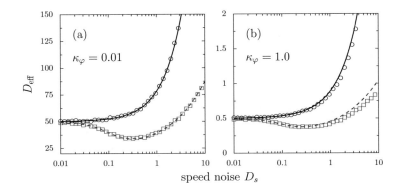

Figure 4.4.: Comparison of D_{eff} given by Eq. (4.23) and (4.24) with numerical simulations for the SET friction function ($\alpha = 1$, $\beta = 50$, circles) and RH friction function ($\alpha = 1$, squares): (a) $\kappa_\varphi = 0.01$; the lines show the approximate result for D_{eff} from Eq. (4.24) (SET – solid line; RH – dashed line); (b) $\kappa_\varphi = 1.0$ the lines show D_{eff} from Eq. (4.23) ($\kappa_v = \alpha = 1$, SET – solid line; RH – dashed line).

For $\kappa_\varphi \ll \kappa_s$ it can be approximated as

$$D_{\text{eff}}^{(2d)} \approx \frac{\langle s^2 \rangle}{2\kappa_\varphi}. \tag{4.24}$$

The effective diffusion coefficient (4.23) is directly dependent on the first and second moment of the speed dynamics. Thus, from our results in the previous section, we can deduce the corresponding behavior of $D_{\text{eff}}^{(2d)}$ with increasing speed noise intensity D_s: e.g. a minimum for the RH-friction and continuous increase for the SET-function.

As shown in Fig. 4.4 the above result (4.24) is confirmed by numerical calculations of the effective diffusion coefficient. Please note that for large D_s systematic deviations appear. The deviations are due to the assumption of exponential decay of speed correlations with the rate κ_s. Here, we used $\kappa_s = \alpha$, which in the case of non-linear propulsion functions is only an approximation.

4.4. Continuous Angular Diffusion

As a second example, we consider continuous angular diffusion due to an (internal) random force $\sqrt{2D_\varphi}\xi_\varphi$ on the angular degree of freedom as discussed in Chapter 3.

The equations of motion for s and φ read

$$\dot{s} = f(s) + \sqrt{2D_s}\xi_s , \tag{4.25}$$

$$\dot{\varphi} = \frac{1}{s}\sqrt{2D_\varphi}\xi_\varphi . \tag{4.26}$$

The Eq. 4.25 corresponds directly to Eq. 4.2, whereas the second equation determines the evolution of the angle of the point particle. The s^{-1} dependence of the angular dynamics reflects the fact that the curvature of the trajectory for a constant force acting in the direction perpendicular to the direction of motion decreases with increasing particle speed. Please note that the angular dynamics diverges for $s \to 0$. This is due to the fact that we are considering a point particle which may instantaneously change its direction of motion according to the applied force at vanishing speed.

This divergence is "unphysical" for real biological agents because a finite-sized, not-moving individual needs always a finite time to turn. We make the following ansatz for the turning dynamics of a stationary agent based on the rotation of an extended object on a substrate

$$I\ddot{\varphi} + \rho\dot{\varphi} = T. \tag{4.27}$$

Here, I being the moment of inertia of the agent, ρ the coefficient of rotational friction and $T \sim F_\varphi$ the torque created by the turning activity of the agent. With the assumption of overdamped angular dynamics, we may neglect the inertia term and obtain

$$\dot{\varphi} = \frac{T}{\rho} = \frac{F_\varphi}{\mu}, \tag{4.28}$$

with $\mu = const. > 0$ defining the finite turning rate of a stationary agent for a given F_φ.

An agent moving with a finite speed $s > 0$ under the influence of a constant force F_φ moves on a circle with a radius r_c determined by s and F_φ. The angular frequency equals the time derivative of the heading angle: $\omega = s/r_c = \dot{\varphi}$, thus the inertial force of rotational motion (centrifugal force) can be rewritten as

$$\omega^2 r_c \mathbf{e}_r = \omega s \mathbf{e}_r = -\dot{\varphi}s\mathbf{e}_\varphi, \tag{4.29}$$

where $\mathbf{e}_r$ is the unit vector in the radial direction perpendicular to the heading direction $\mathbf{e}_h \mathbf{e}_r = 0$ and pointing in the opposite direction to the angular unit vector: $\mathbf{e}_r = -\mathbf{e}_\varphi$. Taking the inertial force, acting on a moving agent, into account modifies Eq. 4.28 to

$$\mu\dot{\varphi} = F_\varphi - \dot{\varphi}s, \tag{4.30}$$

which gives us finally

$$\frac{d\varphi}{dt} = \frac{F_\varphi}{\mu + s}. \tag{4.31}$$

For a time varying force $F_\varphi(t)$, at each point in time t_0, we can define an osculating circle corresponding to the circular agent trajectory for a constant applied force equal to $F_\varphi(t_0)$ (Fig. 3.1).

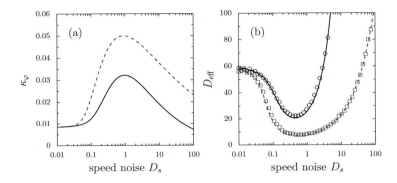

Figure 4.5.: (a) The effective angular diffusion κ_φ (Eq. 4.34) for the SG propulsion function (sold line) and for the RH-propulsion (dashed line) for $\mu = 0.1$ and $D_\varphi = 0.01$. (b) D_{eff} for continuous angular diffusion given by Eq. (4.24) with and corresponding simulations results for the SG-model (simulation – circles; Eq. (4.24) – solid line) and RH friction function (simulation – squares; Eq. (4.24) dashed line). Other parameters values: $\alpha = 1.0$, $D_\varphi = 0.01$.

Assuming that it also holds when F_φ is given by a Gaussian white noise with the intensity D_φ, we obtain the following Langevin equation for the evolution of the angle of an finite-sized agent:

$$\frac{d\varphi}{dt} = \frac{1}{s + \mu}\sqrt{2D_\varphi}\,\xi_\varphi. \tag{4.32}$$

For large speeds s it approaches the corresponding equation of motion of a point-like particle.

The corresponding Fokker-Planck equation for the angular PDF $q(\varphi, t)$ reads

$$\frac{\partial q}{\partial t} = \frac{D_\varphi}{(s + \mu)^2}\frac{\partial^2 q}{\partial^2 \varphi}. \tag{4.33}$$

Whereas the stationary distribution starting from arbitrary initial condition can be easily determined to $q = 1/(2\pi)$, the angular correlation function and in particular the relaxation time $\tau_\varphi = 1/\kappa_\varphi$ depends on the speed s. In other words the evolution of the angle is correlated to the speed and the joint probability density $p(s, \varphi, t|s', \varphi', t')$ cannot be simply decomposed into the product of independent speed and angle PDFs: $p(s, t|s', t')$, $q(\varphi, t|\varphi', t')$ (see also Chapter 3).

We make the same ansatz as in Chapter 3: We consider the case of slow angular dynamics with respect to the speed evolution. Therefore for a given direction of motion φ we may assume that the speed dynamics are given by the stationary speed distribution $p(s)$. In this situation, we approximate the effective angular relaxation rate κ_φ by calculating the expectation value of the pre-factor on the right hand side of Eq. 4.33 with respect to the stationary speed distribution:

$$\kappa_\varphi = \left\langle \frac{D_\varphi}{(s+\mu)^2} \right\rangle = D_\varphi \int_0^\infty ds' \frac{1}{(s'+\mu)^2} p(s'). \tag{4.34}$$

In the case of vanishing speed noise $D_s \to 0$, the stationary speed distribution converges towards a δ-peak at the value of the stationary speed s_0: $p(s) = \delta(s - s_0)$ and the effective angular diffusion reads

$$\kappa_\varphi = \frac{D_\varphi}{(s_0 + \mu)^2} \ . \tag{4.35}$$

The effective coefficient of spatial diffusion can be calculated using (4.24) to

$$D_{\text{eff}}^{(2d)} = \frac{s_0^2}{2\kappa_\varphi} = \frac{s_0^2 (s_0 + \mu)^2}{2 D_\varphi}, \tag{4.36}$$

which for $\mu = 0$ reduces to the result of Mikhailov and Meinköhn (1997) (see also Eq. 3.40) for the diffusion coefficient of self-propelled particles: $D_{\text{eff}}^{(2d)} = s_0^4 / (2 D_\varphi)$.

For stationary speed distributions belonging to the different (non-linear) propulsion functions discussed in this work we were not able to obtain an analytical solutions of the integral determining κ_φ, but it may be easily evaluated by numerical methods. Note that κ_φ depends on the speed noise D_s via its dependence on $p(s)$: with increasing D_s the probability of lower speeds – or fast angular diffusion – increases initially. But with further increase of D_s the width of $p(s)$ increases and the relative frequency of low speeds decreases. Thus κ_φ exhibits a maximum versus D_s (see Fig. 4.5a). This results in a minimum of the spatial diffusion coefficient $D_{\text{eff}}^{(2d)}$ obtained from Eq. 4.24 with respect to D_s as shown in Fig. 4.5b) even for the SG-model.

In summary, we have shown that the behavior of the effective coefficient of spatial diffusion D_{eff} for independent speed and angle dynamics (*run & tumble*) can be be fully understood from the properties of the propulsion function at s_0. The situation changes for continuous angular dynamics where, due to the speed dependent diffusion, we may obtain a minimum of D_{eff} versus D_s even for monotonously increasing speed moments $\langle s^n \rangle$ $(n = 1, 2)$ for sufficiently small D_φ.

5. Optimal Speeds of Animal Motion

There is a long tradition of research on universal scaling laws of metabolic rates of animals with respect to their size (see e.g. Tucker, 1970; Schmidt-Nielsen, 1972; Kram and Taylor, 1990). Many publications have appeared on biomechanics of animal motion and the costs of locomotion (see e.g. Taylor and Heglund, 1982; Fedak et al., 1982; Heglund et al., 1982b,a; Kuo, 2002), as well as on empirical studies of optimal animal speeds minimazing energy consumption (see e.g. Ralston, 1958; Hoyt and Taylor, 1981; Rothe et al., 1987; Sockol et al., 2007).

In this chapter, we will address the question of optimal speeds in animal motion by considering simple metabolic models for the dynamics of an internal energy depot. Within the framework of a general mathematical model, we discuss the existence of optimal speeds, which maximize the distance a biological agent can travel without additional energy input, or the average migration speed of foraging individuals during long distance migration (Sinclair and Arcese, 1995; Alerstam et al., 2003).

We describe an animal as an active agent with an *internal energy* depot. The *internal energy* can be either the energy in a strictly physical meaning, or on a more abstract level any resource critical for the survival of the organism as, such as water for animals in arid environemnts or oxygen for marine mammals.

The general ansatz for the evolution of the internal energy is

$$\frac{\mathrm{d}e}{\mathrm{d}t} = \mathcal{Q} - \mathcal{D} - \mathcal{C} \; . \tag{5.1}$$

Here, $\mathcal{Q}$, $\mathcal{D}$, $\mathcal{C} > 0$ are positive real valued functions describing elementary dynamics of the energy depot: uptake of internal energy, dissipation and conversion into kinetic energy of motion.

Energy uptake: The first term $\mathcal{Q} := \mathcal{Q}(\mathbf{r}, \mathbf{v}, t)$ describes the uptake of energy from the environment ("food intake"). In general, it depends on the spatial distribution of resources (e.g. food patches) and therefore on the position, $\mathbf{r}$, of the agent. It can also depend on the velocity, $\mathbf{v}$, of the agent as, for example, in instances where energy uptake is not possible for a fast moving agent. Finally, it might also depend explicitly on time t (growth and depletion of nutrients).

Energy dissipation: The second term accounts for all internal processes which deplete the internal energy depot without contributing to the locomotion of the agent. This term can be associated with the resting (not moving) metabolic rate of the agent. In the simplest case the dissipation can be assumed as constant, but it might be more reasonable to assume a dependence of the dissipation function on the level of internal energy $\mathcal{D} = \mathcal{D}(e)$, e.g. larger dissipation at larger energy levels.

Energy conversion: The last term, $\mathcal{C} = \mathcal{C}(\mathbf{r}, \mathbf{v})$, accounts for the conversion of energy into kinetic energy of motion. This term represents the energetic costs of locomotion.

The depletion of the energy depot due to locomotion can depend on the position of the agent $\mathbf{r}$ via different costs of locomotion in different environments. The dependence on the velocity of the agent $\mathbf{v}$ includes the obvious dependence on the speed $s = |\mathbf{v}|$ with higher energetic costs of faster motion but might also account for the dependence on the direction of motion $\mathbf{v}/|\mathbf{v}|$ as for example in uphill or downhill motion.

The dynamics of the speed of locomotion s can in principle depend on the level of internal energy depot as, for example, proposed by Schweitzer et al. (1998). But the simple bidirectional coupling of e and s introduced by Schweitzer et. al may result in a strong feedback effect leading to erratic speed/energy dynamics (Erdmann and Göller, 2006). Here, we restrict ourselves to the simplest case of constant speed of motion: $s_0 = const$ with $s_0 > 0$ independent on time t and internal energy e, where $s = s_0 = const.$ may be identified with the preferred velocity of the agent as introduced in Chapter 2. This simplification can be justified in the biological context from the argument of time scales of interest. It is reasonable to assume that there exist physiological or physical limits on the maximum speed of locomotion. In most cases the relaxation time of the speed of an individual to its preferred speed or any value bounded by its maximal speed will be much shorter than the time scales governing the energy depot dynamics, such as the time until depletion of the energy depot or the time needed to move from one source of food to another.

The evolution of the position of an agent moving with constant speed reads

$$\dot{\mathbf{r}} = s_0 \mathbf{e}_h(t), \tag{5.2}$$

with $\mathbf{e}_h(t)$ being the time-dependent unit vector determining the direction of motion of the agent. In the following we will discuss two slightly different models for the evolution of the internal energy.

Assuming constant speed, we can define the energy change per unit distance travelled by the agent l as

$$R(s_0) = \frac{de}{dl} = \frac{de}{dt}\frac{dt}{dl} = \frac{1}{s_0}\frac{de}{dt} = \frac{\mathcal{Q} - \mathcal{D} - \mathcal{C}}{s_0} \tag{5.3}$$

This function relating the energy expenditure to the speed of locomotion (Ralston, 1958) is often used for dissipation rates independent on e (see e.g. Ralston, 1958). In this case, the maximum of $R(s_0)$ corresponds directly to the optimal speed which minimizes energetic costs of locomotion per unit distance travelled.

5.1. Constant Dissipation Rates

First, we consider a simple situation, with a homogeneous distribution of resources, which allow a continuous uptake of energy with a constant rate q_0. The dissipation d_0 is assumed as constant and the speed enters the conversion rate as a parameter with an exponent a multiplied by a conversion factor c:

$$\frac{de}{dt} = q_0 - d_0 - c_0 s_0^a . \tag{5.4}$$

The left hand side of Eq. (5.4) does not depend on t or e and it may be trivally integrated to:

$$e(t) = e(0) + (q_0 - d_0 - cs_0^a)t \ . \tag{5.5}$$

The energy depot increases (decreases) linearly with time for $q_0 > d_0 + cs_0^a$ ($q_0 < d + cs_0^a$). The internal energy e has to stay positive $e \geq 0$. It is biologically and physically reasonable to assume that due to a limited capacity of the depot the energy cannot grow *ad inifinitum*. Therefore, we introduce a corresponding boundary condition for the internal energy. Using the result from Eq. (5.5) we may write the internal energy as:

$$\tilde{e}(t) = \begin{cases} \min(e(t), e_{\max}) & \text{for } q_0 > d_0 + cs_0^a \\ \max(e(t), 0) & \text{for } q_0 < d_0 + cs_0^a \end{cases} . \tag{5.6}$$

For $q_0 > d_0 + cs_0^a$ the internal energy increases until it reaches the capacity of the internal depot $e_{\max}$ and remains constant afterwards, whereas for $q_0 < d_0 + cs_0^a$ it decreases linearly until it reaches $\tilde{e} = 0$

We consider now the case where the individual moves in an environment with insufficient nutrients where $q_0 < d_0 + cs_0^a$. This might correspond to a situation where the individual is displaced by external factors from a nutrient rich environment, or to the situation where the individual itself consumed all the locally available nutrients. In this case, we can combine the uptake rate q_0 and the dissipation rate d_0 to an effective dissipation rate $\tilde{d}_0 = d_0 - q_0 > 0$.

Under the given circumstances we ask the question: Is there an optimal (stationary) speed $s_0 = s_{\text{opt}}$ which maximizes the distance the agent may travel before its energy depot is depleted?

For an individual with full knowledge about the nearest nutrient source which allows effective energy uptake, the best strategy is to move straight in the respective direction. This corresponds to a ballistic motion. In this case, irrespective of the dimensionality of the motion, the maximal distance the individual can cover before its energy depot is depleted is

$$\lambda(s_0) = \frac{s_0 e_{\max}}{\tilde{d}_0 + cs_0^a} \ . \tag{5.7}$$

Here, we set the initial energy value to $e_{\max}$. We assume in (5.7) that the agent moves permanently starting from $t = 0$ until

$$t_{\max} = \frac{e_{\max}}{\tilde{d}_0 + cs_0^a}, \tag{5.8}$$

where its energy depot is depleted.

The above results (Eq. 5.7) gives us the maximal distance, which an individual moving at a speed s_0 may travel in an environment with insufficient nutrients (decreasing e). If food patches with sufficient nutrients are separated by a characteristic distance l an individuals moving at speed s_0 can only survive if l is smaller than $\lambda(s_0)$.

The maximal distance $\lambda(s_0)$ has only a single maximum only if $a > 1$ (see Fig. 5.1a).

By using elementary analysis we can calculate the corresponding optimal speed to:

$$s_{\text{opt}} = \left(\frac{\tilde{d}_0}{(a-1)c} \right)^{\frac{1}{a}} \qquad \text{with} \qquad a > 1 \ . \tag{5.9}$$

The optimal speed decreases with decreasing (increasing) $\tilde{d}_0$ (c) as a power-law and shows a non-monotonous behavior with the exponent a (Fig. 5.1b,c).

Interestingly, all experimentally obtained values for the exponent a reported in the literature are in the range where an optimal speed exists: $1 < a \leq 2$ (see e.g. Heglund et al., 1982b).

We can insert the optimal speed Eq. (5.9) into the distance travelled before the depletion of the energy depot to obtain the limiting value which the individual can travel before starvation irrespective of its speed:

$$\lambda_{\text{max}} = \frac{e_{\text{max}}(a-1)}{a} \frac{(\tilde{d}_0)^{\frac{1}{a}-1}}{(c(a-1))^{\frac{1}{a}}} \tag{5.10}$$

This is the limiting value on the characteristic distance l between nutrient sources only in terms of the metobolic constants. It follows a simple power-law dependence versus $\tilde{d}_0$ and c.

Concluding this section, we apply our results to humans by taking the empirical findings on the metabolic constants for walking humans obtained by Ralston (1958): $d_0 = 2.023\text{J}/(\text{s} \cdot \text{kg})$, $c = 1.3314\text{Js}/(\text{kg} \cdot \text{m}^2)$ and $a = 2$. We can calculate the optimal speed using Eq. 5.9 to

$$s_{\text{opt,human}} = 1.23 \frac{m}{s} \ .$$

This result was already obtained by Ralston in 1958 but using the simple theory we may in principle also calculate the theoretical value of the time before starvation and maximal distance a human can walk (Eq. 5.10). The problem arises with the choice of the initial value for e. Inserting the approximate daily intake of an adult male as initial internal energy ($e(t=0) = 8000\text{kJ} =$) yields

$$t_{\text{max}} \approx 23 \text{ days}, \qquad \lambda_{\text{max}} \approx 2440\text{km}. \tag{5.11}$$

It should be emphasized that Ralston obtained its empirical data from short term experiments. The extremely long t_{max} conflicts with the assumption of continuous activity, as the individuals will not be able to sustain activity for such a long time. This extremely large time and distance suggest that the energy depletion in humans is not the limiting factor and that other factors such as, for example, lack of water are much more important. Unfortunately, until now we were not able to find any scientific studies which would allow us to estimate the costs of locomotion in terms of water needs.

The situation may be different in other species and given empirical data on metabolic coefficients of different species it would be interesting to analyze the theoretical predictions on t_{max} and λ_{max} with actual ecological constraints and observations. We expect that the simple theory is better suited for the description of microorganisms such as bacteria (Mitchell, 2002).

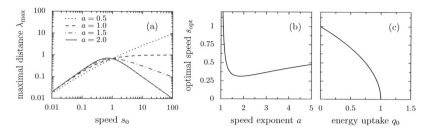

Figure 5.1.: Constant dissipation rate d and constant energy uptake u: **(a)** Maximal distance before energy depot depletion λ_{max} for different speed exponents a (other parameters $q_0 = 0.5$, $e_{max} = c = d = 1.0$);
(b) Optimal speed s_{opt} versus speed exponent a for $d = c = 1.0$ and $q_0 = 0.9$;
(c) Optimal speed s_{opt} versus energy uptake q_0 for $d = c = 1.0$ and $a = 2$.

5.1.1. "Move-or-Forage" — A Simple Model of Migration

In this section we will consider a simple model for migration under the assumption that the biological agent has to sustain an energy level within certain bounds while migrating.

In the preceding section, we did not exclude the possibility that the agent can also take up energy from the environment while moving at an arbitrary speed. For many biological situations this is unrealistic as, for example, terrestrial animals have to slow down if not to stop in order to forage. Thus, we now consider a modified model with two states where the agent can either move with $s = s_0 > 0$ without being able to take up energy, or it can stop to forage ($s = 0$).

Starting with an initial energy e_{max} the agent moves initially and the energy equation reads:

$$\frac{de}{dt}\Big|_{\text{moving}} = -d_0 - cs_0^a, \tag{5.12}$$

thus, the internal energy decreases. As soon as the energy level reaches $e = e_{min}$ the agent stops in order to forage and does so until $e = e_{max}$. During the foraging the energy equation reads:

$$\frac{de}{dt}\Big|_{\text{foraging}} = q_0 - d_0 . \tag{5.13}$$

Here, $q_0 - d_0 > 0$ is the effective energy uptake rate per unit time during foraging. The energy and speed dynamics are periodic in time. The speed is given by a periodic step function, whereas the internal energy follows a saw-tooth function where moving phases correspond to a linear decrease and foraging phases correspond to linear increase of e as shown in Fig. 5.2.

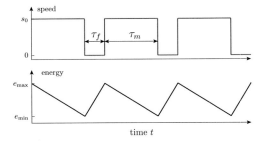

Figure 5.2.: Internal energy and speed versus time for the stationary foraging model with constant dissipation d_0.

The duration of the moving phases is:

$$\tau_m = \frac{\Delta e}{d_0 + c s_0^a} \tag{5.14}$$

with $\Delta e = e_{\max} - e_{\min}$, whereas the duration of the foraging phases reads

$$\tau_f = \frac{\Delta e}{q_0 - d_0}. \tag{5.15}$$

On the one hand, larger speed increase the distance travelled per unit time but on the other hand decrease the moving time τ_m due to faster energy depot depletion. The proportion of time spent moving within one run-and-forage cycle $T = \tau_m + \tau_f$ is τ_m/T and the distance travelled per cycle period T reads

$$\frac{\Delta \lambda}{T} = \frac{s_0 \tau_m}{T} = \frac{s(q_0 - d_0)}{q_0 + c s_0^a}. \tag{5.16}$$

Please note that for the simple model of constant rates $\Delta \lambda / T$ is independent of the energy difference Δe and maximal energy $e_{\max}$. After n-cycles the total distance travelled is $\lambda = n s \tau_m$. Introducing a scaled time $\tilde{t} = nT$ we obtain, in the long time limit ($n \to \infty$ and $T \to 0$), the mean migration speed of the agent:

$$\langle S \rangle = \frac{d\lambda}{d\tilde{t}} = \frac{\Delta \lambda}{T} = \frac{s_0(q_0 - d_0)}{q_0 + c s_0^a}. \tag{5.17}$$

The mean migration speed increases monotonously with q_0 with an asymptotic value of $\langle S \rangle = s_0$ for $q_0 \to \infty$ (see Fig. 5.3a). For $a > 1$ the mean migration speed exhibits a maximum with respect to the speed of motion s_0 (see Fig. 5.3b). We can easily calculate the optimal speed s_{opt} which maximizes $\langle S \rangle$ to

$$s_{\mathrm{opt}} = \left(\frac{q_0}{(a-1)c} \right)^{\frac{1}{a}}. \tag{5.18}$$

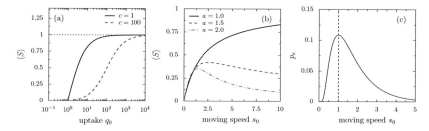

Figure 5.3.: "Move or Forage" model with constant dissipation. **(a)** The mean migration speed $\langle S \rangle$ as a function of the energy uptake q_0 for different values of the conversion factor c. The dotted line represents the asymptote given by the speed during the moving phase $s_0 = 1$. **(b)** The mean migration speed $\langle S \rangle$ versus speed during moving phase s_0 for different values of the exponent a. **(c)** Probability of survival of prey after passing the predator home range ($r = 0.01$, $L = 100$, $u = 1.0$, $d_0 = 0.1$, $a = 2.0$). The dashed line indicates the optimal speed s_{opt}.

Interestingly, the optimal speed does not depend on the constant energy dissipation rate d_0. Thus, the simple model suggests that with respect to maximizing the migration speed there should exist an evolutionary pressure on increasing the foraging rate or decreasing the direct energetic costs of locomotion but not on decreasing the resting metabolic rate.

The idea can be further worked out if we consider the scenario of a migrating prey species e.g. wildebeests, which on their migration route, have to pass the distance L through the homing range of a predator species, e.g. cheetahs or lions (Sinclair and Arcese, 1995). We assume that L is much larger then the typical distance travelled during a foraging cycle $L \gg \Delta\lambda$. The death rate r of the migrating prey species during the passage due to predator attacks is assumed to be constant. The time dependent survival probability for a prey p within the home range of the predator reads:

$$p(t) = e^{-rt}, \qquad (5.19)$$

with $p(0) = 1$. Using $t = L/\langle s \rangle$ we obtain the survival probability of the migrating prey after passing the predator homing range as a function of the moving speed s_0 to:

$$p_e(s_0) = e^{\frac{-rL(q_0 + cs_0^a)}{s_0(q_0 - d_0)}} \quad . \qquad (5.20)$$

The final survival probability exhibits – as expected – a peak at the optimal speed given in Eq. (5.18) (see Fig. 5.3c). This rather simple but reasonable scenario shows how moving at optimal speed provides a direct benefit for the survival of the migrating prey.

Finally, we may insert the optimal speed s_{opt} (5.18) into the expression for the mean migration speed (5.17) to obtain the maximal sustainable migration speed:

$$\langle S \rangle_{\max} = \frac{(q_0 - d_0)(a-1)q_0^{\frac{1}{a}-1}}{a(c(a-1))^{\frac{1}{a}}}, \tag{5.21}$$

Again, the obtained result depends only on metabolic factors, which, at least in principle, can be measured in experiments.

5.2. Energy Dependent Dissipation Rate

Now we consider slightly modified energy dynamics by assuming that the dissipation function is not simply a constant but depends also linearly on e leading to large dissipation rates at higher internal energy levels:

$$\mathcal{D}(e) = -d_0 - d_1 e, \tag{5.22}$$

with $d_1 > 0$ being the proportionality factor.

For continuous, speed-independent energy uptake $q_0 = const.$ the energy equation reads

$$\frac{de}{dt} = q_0 - d_0 - d_1 e - cs_0^a, \tag{5.23}$$

with the time dependent solution

$$e(t) = \left(e_0 + \frac{d_0 + cs_0^a - q_0}{d_1}\right) e^{-d_1 t} - \frac{d_0 + cs_0^a - q_0}{d_1}. \tag{5.24}$$

In the limit $t \to \infty$ the energy e converges asymptotically to a constant level:

$$e_a = \frac{q_0 - d_0 - cs_0^a}{d_1} \tag{5.25}$$

with $e_a > 0$ for $q_0 > d_0 + cs_0^a$. Therefore, in contrast to the simple model studied above, we do not need to introduce an explicit upper limit on $e(t)$. If $q_0 < d_0 + cs_0^a$, the asymptotic value is negative, which requires us to ensure the positivity of the internal energy by using the following definition:

$$\tilde{e}(t) = \max(0, e(t)). \tag{5.26}$$

For a resting individual $s = 0$ under sufficient supply of food, the internal energy approaches exponentially a limiting value of $e_{\max} = (q_0 - d_0)/d_1$. Therefore $e_{\max}$ can be seen as the maximal capacity of the internal energy depot.

For an agent moving at a constant speed s_0 in an environment with sufficient nutrients, the stationary level of internal energy $e_s < e_{\max}$ is given as

$$e_s = \frac{q_0 - d_0 - cs_0^a}{d_1}. \tag{5.27}$$

We consider first the case, where the agent is placed at a time t_0 with an initial energy e_i at a position with no nutrients $u = 0$. Regarding the choice of the initial value of the energy depot, we distinguish two cases:

1. The initial energy $e_i = e_0$ is independent of the agent speed s_0.

2. The initial energy corresponds to the stationary value $e_i = e_s$ of a permanently moving agent with sufficiently large energy uptake $q_0 > d_0 + cs_0^a$.

The first case corresponds to an agent which rests ($s_0 = 0$) at a position with nutrients and then starts to move at a certain point in time with $q_0 = 0$ (e.g. after depletion of the nutrient source). The second case corresponds to the situation of a permanently moving agent, which passes through a region with nutrients for sufficiently large time in order to reach the stationary value of the energy depot e_s and after leaving the region continues its motion without being able to take up additional energy $q_0 = 0$.

The solution for the first case can be directly obtained from Eq. (5.24) by setting $q_0 = 0$,

$$e_1(t) = \left(e_0 + \frac{d_0 + cs_0^a}{d_1} \right) e^{-d_1 t} - \frac{d_0 + cs_0^a}{d_1} , \qquad (5.28)$$

whereas for the second case we obtain

$$e_2(t) = \frac{q_0}{d_1} e^{-d_1 t} - \frac{d_0 + cs_0^a}{d_1} . \qquad (5.29)$$

Here, we set $t_0 = 0$ without loss of generality.

From Eqs. 5.28, 5.29 we can easily determine for the two cases the time $t_{\max}$ when, without additional energy supply, the internal energy of the agent is depleted ($e(\tau) = 0$):

$$t_{\max,1} = \frac{1}{d_1} \ln \left(\frac{e_i d_1}{d_0 + cs_0^a} + 1 \right) , \qquad (5.30)$$

$$t_{\max,2} = \frac{1}{d} \ln \left(\frac{q_0}{d_0 + cs_0^a} \right) . \qquad (5.31)$$

The maximal distance the individual can cover in balistic motion before it "starves" is $\lambda = st_{\max}$. For the two cases we obtain

$$\lambda_1(s_0) = \frac{s}{d_1} \ln \left(\frac{e_i d_1}{d_0 + cs_0^a} + 1 \right) \qquad (5.32)$$

$$\lambda_2(s_0) = \frac{s}{d_1} \ln \left(\frac{q_0}{d_0 + cs_0^a} \right) \qquad (5.33)$$

In order to determine the condition for the existence of optimal speed s_{opt}, which maximizes λ_i we check the existence of an extremum by setting the derivative of λ_i with respect to s_0 to 0. For the first case we obtain

$$\frac{\partial \lambda_1}{\partial s_0} = \frac{1}{d_1} \ln \left(1 + \frac{d_1 e_i}{d_0 + cs_0^a} \right) - \frac{ace_i s_0^a}{(1 + \frac{d_1 e_i}{d_0 + cs_0^a})(d_0 + cs_0^a)^2} = 0 . \qquad (5.34)$$

It is not possible to obtain the root of Eq. 5.34 as an analytical expression, but it can easily be found by standard root-finding algorithms. From the properties of the logarithm function $\ln(1 + x) \le x$ in Eq. 5.32 the existence of a maximum of $\lambda(s_0)$ can be shown for $a > 1$ (Fig. 5.2.1). In fact, it is obvious that in the limit $d_1 \ll d_0$ the optimal speed is well approximated by the result for the simple model with $d_1 = 0$ (5.9).

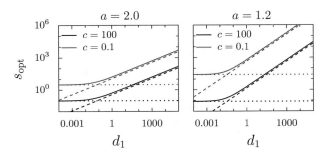

Figure 5.4.: Optimal speed s_{opt} versus the dissipation constant d_1 for $\alpha = 2.0$ (**left**) and $\alpha = 1.2$ (**right**). The solid lines represent the solution found by solving Eq. (5.34) numerically. The dotted lines represents results obtained from Eq. 5.9. The dashed line represent the results form Eq. 5.35. For all curves shown $e_i = 100.0$.

By setting $d_0 = 0$ and assuming $d_1 e_i \gg 1$, with $1 + (d_1 e_i)(d_0 + cs_0^a) \approx (d_1 e_i)(d_0 + cs_0^a)$, we may approximate the solution of Eq. (5.34) to

$$s_{\text{opt},2} \approx \frac{1}{e^1} \left(\frac{d_1 e_i}{c} \right)^{\frac{1}{\alpha}} . \tag{5.35}$$

A comparison of this result with the numerical solution of (5.34) shows that it systematically underestimates the value of s_{opt} due to the approximation made, but gives us the right scaling of s_{opt} with the metabolic parameters (Fig. 5.4).

This can be understood from the fact that at any point $e(t) = e_t$ (for decreasing energy) we may approximate $e(t + dt)$ for small dt by a Taylor expansion with $e_t = const.$. This leads to an effective dissipation rate of the form: $\tilde{d} = d_0 + d_1 e_t = const.$ In the limit of $d_1 e_t \ll d_0$ the constant dissipation dominates and retain the simple model.

In the second case, with initial energy $e_i = e_s$ (Eq. (5.27)) depending on s_0, the derivative reads

$$\frac{\partial \lambda_2}{\partial s_0} = \frac{1}{d_1} \left[\ln \left(\frac{u}{d_0 + cs^a} \right) - \frac{acs^a}{d_0 + cs^a} \right] = 0 \tag{5.36}$$

Again, in general it is not possible to obtain an analytical solution but for $d_0 = 0$ the above equation can be easily solved to

$$s_{\text{opt},2} = \frac{1}{e^1} \left(\frac{q_0}{c} \right)^{\frac{1}{a}} \quad \text{for} \quad d_0 = 0. \tag{5.37}$$

In this case, it can be easily shown from $\partial^2 \lambda_2 / \partial s_0^2 < 0$ that a maximum of λ_2 exists for any positive value of the exponent $a > 0$. This holds also for arbitrary values of d_0 which follows from the analysis of the asymptotic behavior of λ_2 for $s_0 \to 0$ and $s_0 \to \infty$.

5.2.1. Random Walk

We can also analyze the situation of an uninformed individual, which performs a random walk in order to find the next food patch. In this case, the maximal length reached by an agent varies due to the stochastic nature of its motion. But we can calculate the mean square of the maximal displacement $\langle \lambda^2 \rangle$ according to

$$\langle \lambda^2 \rangle (s_0) = 2dD(s_0) \left(t_{\max} - \frac{1}{\kappa} \left(1 - e^{-\kappa t_{\max}} \right) \right) \tag{5.38}$$

Here, d is the dimension of the random walk ($d = 1, 2, 3$), $D(s_0)$ is the spatial diffusion coefficient, which varies for different random walk models and can be calculated from the angular correlation function. κ^{-1} is the characteristic time at which the information about the initial orientation of the agent is lost. As the speed of the individual increases, the time before the depletion decreases $t_{\max} \to 0$. In this limit, Eq. 5.38 reduces to the deterministic solutions 5.32 and 5.33. Therefore we obtain for the random search the same result with respect to the existence of an optimal speed for arbitrary exponents a in the second case and for $a > 1$ in the case where the initial energy does not depend on the speed s_0.

Here, we consider a simple random walk model in two dimensions ($d = 2$). The agent performs a *run & tumble* motion (see also Section 4.3): It moves with a constant speed s_0 and changes its direction of motion at random times. The angles before and after a *tumbling* event are uncorrelated. The time intervals Δt between two consecutive reorientation events are exponentially distributed $P(\Delta t) = \exp(-\kappa t)/\kappa$. The spatial diffusion coefficient for this simple random walk reads

$$D(s_0) = s_0^2/(2\kappa) .$$

In Fig. 5.2.1, we compare the result of Eq 5.38 with numerical simulations of the simple random walk. The numerical results confirm our analytical findings, and show that the framework can be extended to stochastic motion dynamics of individuals in animal search problems.

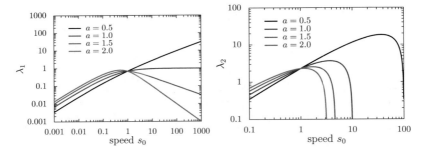

Figure 5.5.: Results for the maximal distance $\lambda_i(s_0)$ for the deterministic motion for the two different cases: initial energy independent of the speed (**left**); initial energy being the stationary energy level of a moving individual (**right**).

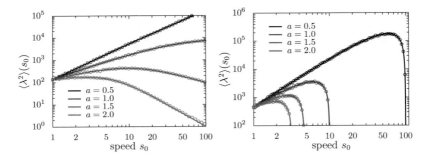

Figure 5.6.: Results for the mean square maximal distance $\langle \lambda^2 \rangle$ for the random search. Comparison of theory (lines) and simulation (symbols) for the two different cases: initial energy independent of the speed (**left**); initial energy being the stationary energy level of a moving individual (**right**)

Part II.

Theory of Collective Dynamics

6. Kinetic Theory of Active Brownian Particles with Velocity Alignment

In this chapter, we consider an active Brownian particle gas with (external) fluctuations and a velocity-alignment interaction. In particular, we will derive a kinetic mean-field theory for the macroscopic observables of the system. In our further analysis, we will focus for simplicity on the stationary, spatially homogeneous mean-field solutions.

The velocity-alignment interaction studied here (Czirok et al., 1996), can be seen as a continous version of the well known Vicsek-model (Vicsek et al., 1995) and reduces for self-propelled particles with constant speed to the polar-alignment model studied by Peruani et al. (2008).

There has been a number of publication on the hydrodynamics of self-propelled particles with velocity alignment and related coarse-grained flocking theories (see, e.g., Toner and Tu, 1998, 1995; Bertin et al., 2006; Simha and Ramaswamy, 2002a,b, or Ramaswamy, 2010 for a recent review). Toner and Tu (1995) constructed mesoscopic equations of motion for the density and velocity fields using symmetry and conservation laws. Recently, Bertin et al. (2006) derived hydrodynamic equations of interacting self-propelled particles by a Boltzmann approach. Here, in contrast to the previous publications, we derive the mean-field equations in a systematic way from the microscopic Langevin equations. We do not assume a constant speed and analyze the impact of the (nonlinear) velocity-dependent friction function on the onset of collective motion. In addition to the density and velocity fields, we consider explicitly the effective temperature field of the active Brownian particle gas.

The derivation of the kinetic theory is based on the formulation of moment equations of the corresponding probability distribution. This approach has been employed by Riethmüller et al. (1997) to analyze the behavior of a quasi one-dimensional granular system. Erdmann (2003) used it to analyze the mean-field of non-interacting active particles with Rayleigh-Helmholtz friction, whereas Strefler (2007) applied the same approach to active Brownian particles with pair-wise attraction/repulsion interaction (Morse potential).

In general, for a system far from equilibrium the probability distribution is not Gaussian and a correct description requires infinitely many moments (see for example Pawula (1967, 1987)). Thus, depending on the detailed model, it may be necessary to neglect higher moments in order to obtain a closure of the system of moment equations. The approximation of a non-Gaussian probability distribution by a finite number of moments may lead to unphysical behavior, such as negative values or artificial oscillations of the (approximated) probability distribution. Therefore, we will compare our analytical results to numerical simulations of the microscopic system.

6.1. General Velocity Alignment Model and Definition of Moments

We consider a system of N active Brownian particles with mass m in d spatial dimensions. The evolution of the particle positions $\mathbf{r}_i$ and velocities $\mathbf{v}_i$ is described by the following set of stochastic differential equations ($i = 1 \ldots N$):

$$\dot{\mathbf{r}}_i = \mathbf{v}_i \tag{6.1}$$

$$m\dot{\mathbf{v}}_i = -\gamma(\mathbf{v}_i)\mathbf{v}_i + \mu(\mathbf{u}_{\varepsilon,i} - \mathbf{v}_i) + \sqrt{2D}\boldsymbol{\xi}_i(t) \tag{6.2}$$

The first term on the right hand side is a friction force, which determines the velocity dynamics of non-interacting particles. In the following sections, we will discuss the collective dynamics of active particles with two different friction functions: the Rayleigh-Helmholtz friction as well as a variant of the Schienbein-Gruler friction. In the following, $m = 1$ is used.

The second term describes the velocity-alignment interaction, with $\mathbf{u}_{\varepsilon,i} = 1/N_{\varepsilon,i} \sum_{j=1}^{N_{\varepsilon,i}} \mathbf{v}_j$ being the mean velocity of particles within a finite radius ε around the focal particle[1] i. The alignment strength $\mu = 1/\tau_a > 0$ determines the relaxation time τ_a of the velocity of the focal particle towards the average velocity of surrounding particles. For solitary particles, with no neighbors, or a system in a perfectly ordered state where all particles move with equal velocity, the alignment force vanishes as $\mathbf{u}_{\varepsilon,i} = \mathbf{v}_i$. On the other hand, for a large number of neighbors $N_{\varepsilon,i} \gg 1$ moving with random velocities (disordered state), the mean velocity vanishes $u_{\varepsilon,i} = 0$ and the velocity alignment force leads to an additional "social" friction $-\mu\mathbf{v}_i$.

Finally, the last term on the right hand side of Eq. 6.2 accounts for (external) fluctuations and is given by a Gaussian random force with intensity D. Here, $\boldsymbol{\xi}(t)$ is a random Cartesian vector with uncorrelated and normal distributed components: $\langle \xi_k \rangle = 0$, $\langle \xi_l(t)\xi_k(t') \rangle = \delta(t - t')\delta_{lk}$.

We will derive the mean-field transport theory for the microscopic dynamics in Eqs. (6.1),(6.2) via the formulation of moment equations of the corresponding probability distribution.

In general, for a probability distribution $p(\mathbf{r}, \mathbf{v}, t)d\mathbf{r}d\mathbf{v}$, which determines the probability to find a particle at time t, at position $\mathbf{r}$ moving with velocity $\mathbf{v}$, we define the n-th moment of the k-component of the velocity vector $v_k = \mathbf{v}\mathbf{e}_k$, with $\mathbf{e}_k$ being a canonical basis unit vector, as

$$M_k^{(n)}(\mathbf{r}, t) = \langle v_k^n \rangle = \frac{1}{\rho} \int d\mathbf{v}\, v_k^n p(\mathbf{r}, \mathbf{v}, t), \quad n > 0. \tag{6.3}$$

The normalization ρ is the zeroth moment which is equivalent to the marginal density:

$$M_0(\mathbf{r}, t) = \rho(\mathbf{r}, t) = \int d\mathbf{v}\, p(\mathbf{r}, \mathbf{v}, t). \tag{6.4}$$

Multiplying the n-th moment of with the density and taking the derivative with respect

---

[1] Please note that the sum includes the velocity of the focal particle $\mathbf{v}_i$. Thus, for a solitary particle $N_\varepsilon = 1$ and $\mathbf{u}_{\varepsilon,i} = \mathbf{v}_i$.

to time, we obtain the dynamics of the moments of velocity:

$$\frac{\partial}{\partial t}(M_0 M_k^{(n)}) = \int d\mathbf{v}\, v_k^n \frac{\partial p}{\partial t}. \tag{6.5}$$

For more than one dimension the above definition can easily be extended to mixed moments as, for example, the covariance with $k \neq l$:

$$M_{kl}^{(nm)}(\mathbf{r}, t) = \langle v_k^n v_l^m \rangle = \frac{1}{\rho} \int d\mathbf{v}\, v_k^n v_l^m p(\mathbf{r}, \mathbf{v}, t) \quad n, m > 0. \tag{6.6}$$

The starting point for the derivation of our macroscopic theory is the N-particle distribution function

$$P(\mathbf{R}, \mathbf{V}, t) = P(\mathbf{r}_1, \ldots, \mathbf{r}_N, \mathbf{v}_1, \ldots, \mathbf{v}_N, t), \tag{6.7}$$

which obeys the following Fokker-Planck equation

$$\frac{\partial}{\partial t} P(\mathbf{R}, \mathbf{V}, t) = -\mathbf{V} \frac{\partial}{\partial \mathbf{R}} P - \frac{\partial}{\partial \mathbf{V}} \left\{ -\gamma(\mathbf{V})\mathbf{V}P + \mu(\mathbf{u}_\varepsilon(\mathbf{R}, \mathbf{V}, t) - \mathbf{V})P \right\}$$
$$+ D \frac{\partial^2}{\partial \mathbf{V}^2} P. \tag{6.8}$$

For identical particles, we can integrate out all particles but a focal one with index i to obtain an effective single particle description with a coupling to the mean velocity field u_ε. For simplicity, we will omit in the following the particle index i.

The single particle Fokker-Planck-Equation reads

$$\frac{\partial p(\mathbf{r}, \mathbf{v}, t)}{\partial t} = -\mathbf{v} \frac{\partial}{\partial \mathbf{r}} p - \frac{\partial}{\partial \mathbf{v}} \{\gamma(\mathbf{v})\mathbf{v} + \mu(\mathbf{u}_\varepsilon - \mathbf{v})\} p + D \frac{\partial^2}{\partial \mathbf{v}^2} p. \tag{6.9}$$

Here, $\mathbf{u}_\varepsilon = \mathbf{u}_\varepsilon(\mathbf{r}, t)$ is the mean-field velocity sensed by the focal particle. In the continuous description we can transform the sum into an integral over the probability distribution:

$$\mathbf{u}_\varepsilon = \frac{1}{\int_{S_\varepsilon} d\mathbf{r}'\, \rho(\mathbf{r}', \mathbf{t})} \int_{S_\varepsilon} d\mathbf{r}' \int d\mathbf{v}' \mathbf{v}' p(\mathbf{r}', \mathbf{v}', t). \tag{6.10}$$

Here, S_ε represents the spatial neighbourhood of the focal particle, which is defined via a metric distance: $\mathbf{r}' \in S_\varepsilon$ if $|\mathbf{r}' - \mathbf{r}| < \varepsilon$.

6.2. Rayleigh-Helmholtz Friction

6.2.1. One-dimensional Case

First, we consider a system of Active Brownian Particles with the so-called Rayleigh Helmholtz friction in one spatial dimension ($d = 1$). The equations of motion read

$$\dot{x}_i = v_{x,i} \tag{6.11}$$
$$\dot{v}_{x,i} = (\alpha - \beta v_i^2) v_{x,i} + \mu(u_{\varepsilon,i} - v_{x,i}) + \sqrt{2D}\xi_i(t). \tag{6.12}$$

The Fokker-Planck-Equation for a single particle in the mean velocity field u_ε reads

$$\frac{\partial p}{\partial t} = -v_x \frac{\partial}{\partial x} p - \frac{\partial}{\partial v_x} \{(\alpha - \beta v_x^2)v_x + \mu(u_\varepsilon - v_x)\}p + D \frac{\partial^2}{\partial v_x^2} p. \qquad (6.13)$$

Inserting Eq. (6.13) in Eq. (6.5) and using $\lim_{v_x \to \pm\infty} p(x, v_x, t) = 0$, we can partially integrate the terms with partial derivatives with respect to v_x to obtain:

$$\frac{\partial}{\partial t}(M_0 M_n) = -\frac{\partial}{\partial x}\rho \langle v_x^{n+1}\rangle + n\,\rho\left[\alpha\,\langle v_x^n\rangle - \beta\,\langle v_x^{n+2}\rangle + \mu(u_\varepsilon\langle v_x^{n-1}\rangle - \langle v_x^n\rangle)\right]$$
$$+ n\,(n-1)\,D\,\rho\,\langle v_x^{n-2}\rangle. \qquad (6.14)$$

Please note that the evolution of the n-th moment depends also on the $(n+2)$-th moment which leads to an infinite hierarchy of moment equations. We will keep this in mind while proceeding with the derivation of our mean-field equations up to the second order $n = 2$. We rewrite the velocity of the focal particle as a sum of the local velocity field $u(x, t)$ plus some deviation δv_x: $v_x = u + \delta v_x$. Furthermore, we assume $\langle \delta v_x^a \rangle = 0$ for odd exponents a $(a = 1, 3, 5, \ldots)$. Thus, we obtain for the moments (up to $a = 4$):

$$\langle v_x \rangle = u \,, \qquad (6.15a)$$

$$\langle v_x^2 \rangle = u^2 + T \,, \qquad (6.15b)$$

$$\langle v_x^3 \rangle = u^3 + 3\,u\,T \,, \qquad (6.15c)$$

$$\langle v_x^4 \rangle = u^4 + 6\,u^2\,T + T^2 + \theta \,. \qquad (6.15d)$$

Here, T is the mean squared velocity deviation $T = \langle \delta v_x^2 \rangle$, which we will refer to as the temperature of the active particle gas, whereas θ is the average of the mean squared temperature fluctuations (Erdmann, 2003) defined as

$$\theta = \langle \left((v_x - u_x)^2 - T \right)^2 \rangle = \langle \delta v^4 \rangle - T^2. \qquad (6.16)$$

Now we can insert the Eqs. (6.15) in Eq. (6.14). Considering the dynamics up to $n = 2$, after some calculus, we arrive at a set of three coupled partial differential equations for the evolution of the density $\rho(x, t)$, the mean velocity field $u(x, t)$, and the temperature field $T(x, t)$:

$$\frac{\partial}{\partial t}\rho = -\frac{\partial}{\partial x}(\rho\,u) \qquad (6.17a)$$

$$\frac{\partial u}{\partial t} + u\,\frac{\partial}{\partial x}u = \alpha\,u - \beta\,u\,\left(u^2 + 3T\right) + \mu(u_\varepsilon - u) - \frac{\partial T}{\partial x} - \frac{T}{\rho}\frac{\partial \rho}{\partial x} \qquad (6.17b)$$

$$\frac{1}{2}\left(\frac{\partial T}{\partial t} + u\,\frac{\partial T}{\partial x}\right) = (\alpha - \mu)\,T - \beta\,T\left(3u^2 + T\right) - \beta\,\theta + D - T\frac{\partial u}{\partial x} \qquad (6.17c)$$

For an isotropic, spatially homogeneous system with vanishing gradients in the mean velocity field u and mean temperature field T the above system of equations simplifies to

$$\frac{du}{dt} = \alpha\, u - \beta\, u\, \left(u^2 + 3T\right) \tag{6.18a}$$

$$\frac{1}{2}\frac{dT}{dt} = (\alpha - \mu)\, T - \beta\, T\, \left(3u^2 + T\right) - \beta\theta + D. \tag{6.18b}$$

The above sets of equations for the central moments Eqs. (6.17) and Eqs. (6.18) are not self-consistent as the right hand side of the temperature equations depends on the temperature fluctuations $\theta = \theta(x, t)$. By proceeding to higher orders $n > 2$ it is possible to obtain an equation for θ but it will again depend on a higher order central moment. Here, in order to obtain a closed system of equations, we neglect the temperature fluctuations by setting $\theta = 0$ in Eq. (6.16), which is a reasonable assumptions at small noise intensities.

With this approximation, the above differential equations for the homogeneous system constitute a two-dimensional dynamical system, with 6 fixed points (stationary solutions) in the (u, T) phase space, which can be analyzed by means of linear stability analysis.

The stationary solutions $(du/dt = dT/dt = 0)$ for u and T read:

$$u_{1,2} = 0, \tag{6.19a}$$

$$T_{1,2} = \frac{\alpha - \mu \pm \sqrt{(\alpha - \mu)^2 + 4\beta D}}{2\beta} \tag{6.19b}$$

$$u_{3,4} = \pm\frac{\sqrt{10\alpha - 3\left(\mu - \sqrt{(2\alpha + \mu)^2 - 32\beta D}\right)}}{4\sqrt{\beta}} \tag{6.19c}$$

$$T_3 = T_4 = \frac{2\alpha + \mu - \sqrt{(2\alpha + \mu)^2 - 32\beta D}}{16\beta} \tag{6.19d}$$

$$u_{5,6} = \pm\frac{\sqrt{10\alpha - 3\left(\mu + \sqrt{(2\alpha + \mu)^2 - 32\beta D}\right)}}{4\sqrt{\beta}} \tag{6.19e}$$

$$T_5 = T_6 = \frac{2\alpha + \mu + \sqrt{(2\alpha + \mu)^2 - 32\beta D}}{16\beta}. \tag{6.19f}$$

The kinetic temperature T_j has to be positive, therefore, for $u = 0$, T_1 (positive square root) is the only physically reasonable solution.

The first solution with vanishing mean velocity $u = 0$ describes a disordered phase. For $D = 0$, the temperature $T = \frac{\alpha - \mu}{\beta} = v_0^2$ equals the square of the stationary velocity of individual particles. The kinetic energy of all particles completely consists of fluctuations and no global translational motion occurs.

The second pair of solutions corresponds to translational modes which are stable below a critical noise intensity. The two solutions correspond to translational motion with positive or negative velocity u, thus, to a collective motion of the particles to the left or right. Without noise, $T_{3,4} = 0$, and the stationary mean velocity reduces to $u_{3,4} = \pm\sqrt{\alpha/\beta}$. Increasing the noise raises the kinetic temperature, and results in a decrease of the mean speed $|u|$.

The last solution pair describes unstable modes, for which with increasing noise intensity

D the temperature decreases and the mean speed increases.

For low velocity alignment strength $\mu < 2\alpha/3$, the disordered phase is always a stable solution; for $\mu > 2\alpha/3$, the linear stability analysis of the mean-field equations predicts the existence of a critical noise intensity

$$D_{d,\mathrm{crit}} = \frac{\alpha(3\mu - 2\alpha)}{9\beta}, \qquad (6.20)$$

which determines the stability of the disordered solution. Starting from large noise intensities where the disordered solution is stable and decreasing the noise below $D_{d,\mathrm{crit}}$, we observe a pitchfork-bifurcation, and the disordered phase becomes unstable. Depending on the value of μ, the pitchfork-bifurcation is either sub- or super-critical. For $\mu < 10\alpha/3$, the disordered solution becomes unstable through a collision with the two unstable translational solutions, whereas for $\mu > 10\alpha/3$ no unstable translational solutions exist and the disordered solution becomes unstable directly through the appearance of the two stable translational solutions (see Fig. 6.1). Thus, for $\mu > 2\alpha/3$ and $D < D_{d,\mathrm{crit}}$, only the translational solutions $u_{3,4}$ are stable.

For $\mu < 10\alpha/3$ there exists a second critical noise intensity which determines the stability of the ordered phase (translational solutions, $u \neq 0$). Above the critical noise intensity

$$D_{o,\mathrm{crit}} = \frac{(2\alpha + \mu)^2}{32\beta} \qquad (6.21)$$

all translational solutions become unstable through a saddle-node bifurcation (Fig. 6.1 a,b).

The different possible regimes, with respect to the stability of the ordered and disordered solutions, can be summerized in the following way:

$0 < \mu < \frac{2\alpha}{3}$	$D < D_{o,\mathrm{crit}}$	bistability (ordered + disordered solutions)
	$D > D_{o,\mathrm{crit}}$	only disordered solution stable
$\frac{2\alpha}{3} < \mu < \frac{10\alpha}{3}$	$D < D_{d,\mathrm{crit}}$	only ordered solution stable
	$D_{d,\mathrm{crit}} < D < D_{o,\mathrm{crit}}$	bistability (ordered + disordered solutions)
	$D > D_{o,\mathrm{crit}}$	only disordered solution stable
$\mu > \frac{10\alpha}{3}$	$D < D_{d,\mathrm{crit}}$	only ordered solution stable
	$D > D_{d,\mathrm{crit}}$	only disordered solution stable

In order to test our analytical results, we performed numerical simulations of the microscopic model with periodic boundary condition. Due to the symmetry of the translational solutions $u_3 = -u_4$, we distinguish the disordered phase and the ordered (translational) phase by measuring the global mean speed in our simulations:

$$\langle |u| \rangle = \left\langle \left| \frac{1}{N} \sum_{i=1}^{N} v_i \right| \right\rangle. \qquad (6.22)$$

Here, $\langle \cdot \rangle$ denotes temporal average after the system has reached a stationary state. In order to analyze the stability of the (dis)ordered phase, the simulations were performed with two different initial conditions: perfectly ordered ($u(t = 0) = u_3(D = 0)$) and perfectly disordered state ($u(t = 0) = 0$). Each simulation was run until the system

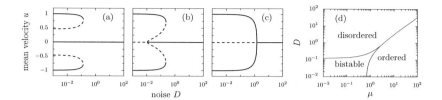

Figure 6.1.: Bifurcation diagram of the mean velocity u vs noise intensity D as predicted from the mean-field theory for different velocity alignment strengths ($\alpha = \beta = 1$) : **(a)** $\mu < 2\alpha/3$ ($\mu = 0.1$), **(b)** $2\alpha/3 < \mu < 10\alpha/3$ ($\mu = 0.7$) and **(c)** $\mu > 10\alpha/3$ ($\mu = 5.0$). **(d)** Phase diagram with respect to velocity alignment μ and noise intensity D.

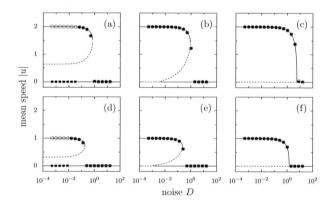

Figure 6.2.: Comparison of the stationary solution obtained from simulation with theoretical prediction from the mean-field theory for different velocity alignment strengths: $\mu = 0.4$ **(a,d)**, $\mu = 0.67$ **(b,e)** and $\mu = 5.0$ **(c,f)** and different friction coefficients: $\beta = 0.25$ **(a,b,c)** and $\beta = 1.0$ **(d,e,f)**. Other parameters used: particle number $N = 8192$, simulation domain $L = 500$, velocity alignment range $\epsilon = 50$ and $\alpha = 1.0$. The initial conditions were either the disordered state (filled squares) or the ordered state (circles). Solid (dashed) lines show the stable (unstable) stationary solutions of the mean-field equations.

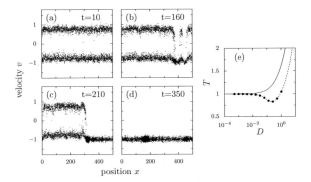

Figure 6.3.: **(a-d)** Simulation snapshots showing the breakdown of the disordered solution. Each point corresponds to a single particle. Starting from a perfectly disordered initial state (**(a)**, $t = 10$), we observe a local symmetry breaking at $x \approx 400$ (**(b)**, $t = 160$) leading to a local velocity alignment of particles. The region of alignment spreads through the system (**(c)**, $t = 210$) until a stationary ordered state is reached (**(d)**, $t = 350$); Simulation parameters: $\alpha = 1.0$, $\beta = 1.0$, $\mu = 0.4$, $D = 0.01$, $N = 4096$, $L = 500$, $\epsilon = 10$. **(e)** Stationary temperature for non-interacting particles ($\mu = 0$) vs. noise intensity D obtained from the mean-field theory (solid line), from direct calculation of the second moment Eq. (6.23) (dashed line) and from numerical simulations (symbols) for $\alpha = \beta = 1$.

reached a stationary state, but at least for $t = 2000$ time units with a numerical time step $\Delta t = 0.01$.

The stationary speeds of the ordered phase versus noise intensity obtained from numerical simulations with an ordered initial condition are in a good agreement with the theoretical predictions from the mean-field theory. Whereas simulations with a disordered initial condition reveal an unexpected instability of the disordered solution. At finite μ the numerical simulations show that at intermediate D the disordered solution $u = 0$ becomes unstable via a spontaneous symmetry breaking as shown in Figs. 6.2 and 6.3a-d, which is not predicted by the mean-field theory.

We reproduced the instability of the disordered solution in numerical simulations for intermediate noise strengths for different particle numbers ($N = 4096, 8192, 16384$). This suggests that it cannot be simply dismissed as a pure finite size effect. Other possible sources for this discrepancy might be the restriction to spatially homogeneous solutions of the mean-field equations, which neglects density fluctuations, or the closure of the moment equation hierarchy by neglecting temperature fluctuations θ.

In order to rule out the first possibility, we have analyzed the case of global coupling $\varepsilon > L$ corresponding to a perfectly homogeneous system. In this case, the disordered solution breaks down as well at intermediate D by a spontaneous (global) symmetry breaking.

Therefore, we conclude that the instability of the disordered mode can be associated with higher order fluctuations. This is also consistent with the agreement of the theoretical result with numerics at low D, where temperature fluctuations are very small. In fact, for non-interacting particles with Rayleigh-Helmholtz friction, we can calculate directly the second moment from the velocity distribution in Erdmann et al. (2000). For $\beta = 1$ it reads

$$\langle v^2 \rangle = \frac{\alpha}{2} \left(1 + \frac{\pi\sqrt{2}I_{-\frac{3}{4}}\left(\frac{\alpha^2}{8D}\right) - K_{-\frac{3}{4}}\left(\frac{\alpha^2}{8D}\right)}{K_{\frac{1}{4}}\left(\frac{\alpha^2}{8D}\right) + \pi\sqrt{2}I_{\frac{1}{4}}\left(\frac{\alpha^2}{8D}\right)} \right), \tag{6.23}$$

with I_n and K_n being the modified Bessel functions of the first and second kind respectively. The result of Eq. (6.23) corresponds directly to the temperature T for the disordered state in the limit $\mu = 0$. Due to the non-linearity of the friction function, the temperature does not increase monotonically with D as predicted by the mean-field theory but exhibits a minimum at intermediate noise intensities as shown in Fig. 6.3e. It can be seen from Eq. (6.18) that the explicit consideration of finite temperature fluctuations ($\theta > 0$) leads to a decrease in T, which is consistent with Eq. (6.23). This in turn decreases the stability of the disordered state. The extension of the mean-field theory to higher orders would account for this effect at the expense of the analytical tractability of the mean-field solutions.

6.2.2. Two-dimensional Case

Here, we will show how the above approach can be extended to higher dimensional systems by deriving the kinetic equations for a gas of particles with Rayleigh-Helmholtz friction in two spatial dimensions ($d = 2$). The equations of motion are (6.1) and (6.2) with $\mathbf{r}_i = (x_i, y_i)$ and $\mathbf{v}_i = (v_{x,i}, v_{y,i})$ being the two dimensional position and velocity vectors, respectively. The Rayleigh-Helmholtz friction function in two dimensions reads

$$\gamma(\mathbf{v}) = -\alpha + \beta\mathbf{v}^2, \tag{6.24}$$

and we may write the single particle Fokker-Planck equation as

$$\frac{\partial p}{\partial t} = -\mathbf{v}\frac{\partial p}{\partial \mathbf{r}} - \frac{\partial}{\partial \mathbf{v}}\{-(\alpha - \beta\mathbf{v}^2)\mathbf{v}p + \mu(\mathbf{u}_\varepsilon - \mathbf{v})p\} + D\frac{\partial^2 p}{\partial \mathbf{v}^2}. \tag{6.25}$$

In analogy to the one-dimensional case ($d = 1$) we insert Eq. (6.25) into Eq. (6.5) and obtain

$$\frac{\partial}{\partial t}(M_0 M_x^{(n)}) = \int v_x^n \left\{ -v_x\frac{\partial}{\partial x} - v_y\frac{\partial}{\partial y} - \frac{\partial}{\partial v_x}\left[-(\alpha - \beta\mathbf{v}^2)v_x + \mu(u_{\varepsilon,x} - v_x) \right] \right.$$

$$\left. - \frac{\partial}{\partial v_y}\left[-(\alpha - \beta\mathbf{v}^2)v_y + \mu(u_{\varepsilon,y} - v_y) \right] + D\left[\frac{\partial^2}{\partial v_x^2} + \frac{\partial^2}{\partial v_y^2} \right] \right\} p\, d\mathbf{v}. \tag{6.26}$$

Since we assume that the probability distribution approaches zero at infinity $\lim_{v_k \to \pm\infty} P = 0$, the terms with derivatives with respect to v_y vanish. The terms with derivatives with respect to v_x can be partially integrated. Therefore, we obtain

$$\frac{\partial}{\partial t}(M_0 M_x^{(n)}) = -\frac{\partial}{\partial x}\left(\rho\left\langle v_x^{n+1}\right\rangle\right) - \frac{\partial}{\partial y}\left(\rho\left\langle v_x^n\, v_y\right\rangle\right)$$
$$+ n\,\alpha\,\rho\left\langle v_x^n\right\rangle - \beta\,\rho\left(\left\langle v_x^{n+2}\right\rangle + \left\langle v_x^n\, v_y^2\right\rangle\right)$$
$$+ n\,(n-1)\,D\,\rho\left\langle v_x^{n-2}\right\rangle. \tag{6.27}$$

We make again the assumption that the velocity $\mathbf{v} = (v_x, v_y)$ is given by the mean-field velocity $\mathbf{u} = (u_x, u_y)$ plus some deviation vector $\delta\mathbf{v} = (\delta v_x, \delta v_y)$ with $\langle \delta v_k^a \rangle = 0$ for odd a. We define the components of the effective temperature vector $\mathbf{T} = (T_x, T_y)$ as the squared deviation of the k-th velocity vector: $T_k := \langle \delta v_k^2 \rangle$. Thus, the moments $\langle v_k^n \rangle$ correspond directly to the moments obtained for $d = 1$:

$$\langle v_k \rangle = u_k \tag{6.28a}$$
$$\langle v_k^2 \rangle = u_k^2 + T_k \tag{6.28b}$$
$$\langle v_k^3 \rangle = u_k^3 + 3\,u_k\,T_x \tag{6.28c}$$
$$\langle v_k^4 \rangle = u_k^4 + 6\,u_k^2\,T_k + T_k^2 + \theta_k, \tag{6.28d}$$

with $k = x, y$ and θ_k being the temperature fluctuations in k-direction defined in (6.16).

Under the assumption of independent deviations in x and y-direction

$$\langle \delta v_x^n \delta v_y^m \rangle = \langle \delta v_x^n \rangle \langle \delta v_y^m \rangle$$

we obtain following expressions for the mixed moments $\langle v_x^n v_y^m \rangle$:

$$\langle v_x\, v_y \rangle = u_x\, u_y \tag{6.29a}$$
$$\langle v_x\, v_y^2 \rangle = u_x\, u_y^2 + u_x\, T_y \tag{6.29b}$$
$$\langle v_x^2\, v_y \rangle = u_x^2\, u_y + u_y\, T_x \tag{6.29c}$$
$$\langle v_x^2\, v_y^2 \rangle = u_x^2\, u_y^2 + u_x^2\, T_y + u_y^2\, T_x + T_x\, T_y. \tag{6.29d}$$

We insert these expressions in our equation for the moment dynamics Eq. (6.27) and, by carrying out the calculations up to the second order, we arrive at a set of differential equations

$$\frac{\partial}{\partial t}\rho = -\nabla_{\mathbf{r}}\left(\rho\mathbf{u}\right), \tag{6.30a}$$

$$\frac{\partial u_x}{\partial t} + \mathbf{u}\nabla_{\mathbf{r}}u_x = \alpha u_x - \beta u_x\left(\mathbf{u}^2 + 3T_x + T_y\right) + \mu(u_{\varepsilon,x} - u_x)$$
$$- \frac{\partial T_x}{\partial x} - \frac{T_x}{\rho}\frac{\partial\rho}{\partial x}, \tag{6.30b}$$

$$\frac{1}{2}\left(\frac{\partial T_x}{\partial t} + \mathbf{u}\nabla_{\mathbf{r}}T_x\right) = (\alpha - \mu)T_x - \beta T_x\left(\mathbf{u}^2 + 2u_x^2 + T_x + T_y\right) - \beta\theta_x$$
$$+ D - T_x\frac{\partial u_x}{\partial x}. \tag{6.30c}$$

Here, we have only given the equation for x-components of the mean velocity and temperature; the corresponding equations for the y-component can be directly obtained by interchanging the x and y indices. For a spatially homogeneous system, the above equations simplify to ordinary differential equations for the mean-field velocity and temperature (x-component):

$$\frac{du_x}{dt} = \alpha u_x - \beta u_x \left(\mathbf{u}^2 + 3T_x + T_y \right), \tag{6.31a}$$

$$\frac{du_y}{dt} = \alpha u_y - \beta u_y \left(\mathbf{u}^2 + T_x + 3T_y \right), \tag{6.31b}$$

$$\frac{1}{2}\frac{dT_x}{dt} = (\alpha - \mu)T_x - \beta T_x \left(\mathbf{u}^2 + 2u_x^2 + T_x + T_y \right) - \beta\theta_x + D, \tag{6.31c}$$

$$\frac{1}{2}\frac{dT_y}{dt} = (\alpha - \mu)T_y - \beta T_y \left(\mathbf{u}^2 + 2u_y^2 + T_x + T_y \right) - \beta\theta_y + D. \tag{6.31d}$$

By setting $\theta_x = \theta_y = 0$, we obtain a self-consistent set of ODE's.

We can further simplify the above set of equations by choosing a reference frame where $u_x = u_{\parallel} = u$ corresponds to the mean-field velocity, whereas the orthogonal component vanishes $u_y = u_{\perp} = 0$:

$$\frac{du}{dt} = \alpha u - \beta u \left(u^2 + 3T_{\parallel} + T_{\perp} \right) \tag{6.32a}$$

$$\frac{1}{2}\frac{dT_{\parallel}}{dt} = (\alpha - \mu)T_{\parallel} - \beta T_{\parallel} \left(3u^2 + T_{\parallel} + T_{\perp} \right) + D \tag{6.32b}$$

$$\frac{1}{2}\frac{dT_{\perp}}{dt} = (\alpha - \mu)T_{\perp} - \beta T_{\perp} \left(u^2 + T_{\parallel} + T_{\perp} \right) + D \tag{6.32c}$$

with $T_{\parallel}$ and $T_{\perp}$ being the temperature components parallel and perpendicular to the mean-field direction of motion. In the stationary disordered state ($u = 0$), the temperature components can be easily calculated from (6.32) and the corresponding solution reads

$$u_1 = 0, \tag{6.33a}$$

$$T_{\parallel,1} = T_{\perp,1} = T_1 = \frac{\alpha - \mu + \sqrt{(\alpha - \mu)^2 + 8\beta D}}{4\beta}. \tag{6.33b}$$

In the case of vanishing noise $D = 0$, the ordered solution can be immediately obtained to $u = \sqrt{\alpha/\beta}$ and $T_{\parallel} = T_{\perp} = 0$. For $D > 0$, it is evident that the temperature component parallel to the direction of motion is smaller than the perpendicular one: $T_{\parallel} < T_{\perp}$.

For the general ordered state, with $u > 0$ and $D > 0$, we were so far not able to obtain explicit stationary solution for u, $T_{\parallel}$ and $T_{\perp}$ of the above ODE system (6.32) but the stable and unstable solutions can be determined by a numerical continuation methods as, for example, provided by the numerical software XPPAUT/AUTO97 (Doedel, 1981; Ermentrout, 2002).

A possible ansatz to find an explicit solution is the reduction of the dimensions in the problem: We use the fact that at a fixed time t we may always find a coordinate frame where $u_x = u_y = \tilde{u}$. In this coordinate frame we obtain due to the symmetry of the involved equations $T_x = T_y = \tilde{T}$. With this ansatz we reduce the original system (6.31)

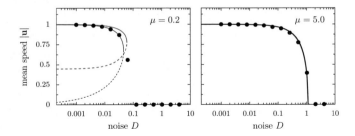

Figure 6.4.: Comparison of the mean-field speed $|u|$ obtained from Langevin simulations (symbols) of the RH-model in two spatial dimensions at high density with the results of the mean-field theory for the homogeneous case. The black lines represent the solutions obtained from the full system of mean-field ODE's. The red lines represent the mean-field solutions from the reduced system. The stable solutions are shown as solid lines, whereas dashed lines indicate the unstable solution. The simulations were performed with periodic boundary condition and with the disordered state as initial condition. Other parameters: $\alpha = 1$, $\beta = 1$, $L = 200$, $\varepsilon = 20$, $N = 4096$.

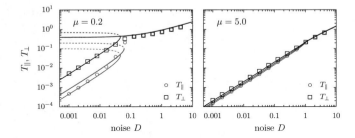

Figure 6.5.: Numerical results for $T_\parallel$ (blue circles) and $T_\perp$ (black squares) obtained from Langevin simulations of the RH-model together with results of the mean-field theory. The solid (dashed) lines represent stable (unstable) solutions. The solutions of the full system for $T_\parallel$ ($T_\perp$) are indicated by blue (black) lines. The red lines show the mean-field solutions from the reduced system. The simulations were performed with periodic boundary condition and with the disordered state as initial condition. Other parameters: $\alpha = 1$, $\beta = 1$, $L = 200$, $\varepsilon = 20$, $N = 4096$.

from a four dimensional system of ODE's to a two dimensional system in $\tilde{u}$ and $\tilde{T}$:

$$\frac{d\tilde{u}}{dt} = \alpha\tilde{u} - \beta\tilde{u}\left(2\tilde{u}^2 + 4\tilde{T}\right) \tag{6.34a}$$

$$\frac{1}{2}\frac{d\tilde{T}}{dt} = (\alpha - \mu)\tilde{T} - \beta\tilde{T}\left(4\tilde{u}^2 + 2\tilde{T}\right) + D \tag{6.34b}$$

This gives us a system of equations similar to the one-dimensional case, where we can analytically determine the stationary solutions for $d\tilde{u}/dt = d\tilde{T}/dt = 0$ to:

$$\tilde{u}_{1,2} = 0, \tag{6.35a}$$

$$\tilde{T}_{1,2} = \frac{\alpha - \mu \pm \sqrt{(\alpha - \mu)^2 + 8\beta D}}{4\beta}, \tag{6.35b}$$

$$\tilde{u}_{3,4} = \pm\frac{\sqrt{2\alpha - \mu + \sqrt{(\alpha + \mu)^2 - 24\beta D}}}{\sqrt{6\beta}}, \tag{6.35c}$$

$$\tilde{T}_3 = T_4 = \frac{\alpha + \mu - \sqrt{(\alpha + \mu)^2 - 24\beta D}}{12\beta}, \tag{6.35d}$$

$$\tilde{u}_{5,6} = \pm\frac{\sqrt{2\alpha - \mu - \sqrt{(\alpha + \mu)^2 - 24\beta D}}}{\sqrt{6\beta}}, \tag{6.35e}$$

$$\tilde{T}_5 = T_6 = \frac{\alpha + \mu + \sqrt{(\alpha + \mu)^2 - 24\beta D}}{12\beta}. \tag{6.35f}$$

The type of the different solutions is completely analogous to the one-dimensional system:

$$\tilde{u}_{1,2}, \tilde{T}_1 \quad : \quad \text{disordered solution}$$
$$\tilde{u}_{3,4}, \tilde{T}_{3,4} \quad : \quad \text{stable ordered solutions}$$
$$\tilde{u}_{5,6}, \tilde{T}_{5,6} \quad : \quad \text{unstable ordered solutions}$$

The structure of the bifurcation diagram of the reduced system (6.34) for $d = 2$ is the same as for $d = 1$ with different critical noise intensities determining the stability of the disordered and ordered solutions:

$$D_{d,\text{crit}}^{(2d)} = \frac{\alpha(2\mu - \alpha)}{8\beta},$$

$$D_{o,\text{crit}}^{(2d)} = \frac{(\alpha + \mu)^2}{24\beta}.$$

The different regimes obtained from the reduced system can be summerized as:

$0 < \mu < \frac{\alpha}{2}$	$D < D_{o,\text{crit}}$	bistability (ordered + disordered solutions)
	$D > D_{o,\text{crit}}$	only disordered solution stable
$\frac{\alpha}{2} < \mu < 2\alpha$	$D < D_{d,\text{crit}}$	only ordered solution stable)
	$D_{d,\text{crit}} < D < D_{o,\text{crit}}$	bistability (ordered + disordered solutions)
	$D > D_{o,\text{crit}}$	only disordered solution stable
$\mu > 2\alpha$	$D < D_{d,\text{crit}}$	only ordered solution stable
	$D > D_{d,\text{crit}}$	only disordered solution stable

The comparison of the stationary solutions of the reduced systems with the corresponding solutions of the full systems obtained with XPPAUT/AUTO reveals differences at low velocity-alignment strengths μ (Fig. 6.4,6.5). The velocity of the stable ordered solution of the full system decreases more strongly and exhibits an earlier breakdown with increasing D. Furthermore, from the position of the disordered branch, it can be deduced that the basin of attraction of the ordered state for the full system for low D is larger than for the reduced two dimensional system. At large μ the differences between the two types of mean-field solution vanish and the reduced system gives a good approximation as shown in Fig. 6.4. The reason for the discrepancy between the two mean-field solutions at low μ is due to the fact that the reduction of the system dimensions throws away all informations about the assymmetry of temperature components parallel and perpendicular to the mean velocity. At large μ the evolution of the temperature coefficients is dominated by the $-\mu T_k$ term and may in a crude approximation simply be assumed as linear for both components, so that the assymmetry in the temperature components can be neglected 6.5.

In general, without knowing the temperatures $T_\parallel$ and $T_\perp$, the mean speed as the order parameter can be written as

$$|u| = \sqrt{v_0^2 - 3T_\parallel - T_\perp} \qquad (6.36)$$

here we used $v_0^2 = \alpha/\beta$. In the limit of large μ close to the critical noise, where $\alpha, \beta \ll \mu, D$, we may approximate the temperature as $T_\parallel = T\perp = T = D/\mu$ and we obtain a simple expression for the ordered state

$$|u| = \sqrt{v_0^2 - \frac{4D}{\mu}}. \qquad (6.37)$$

In this limit, the critical noise can be approximated as $D_{d,\mathrm{crit}} \approx v_0^2\mu/4 = D_{\mathrm{crit}}$ and the above equation may be rewritten as:

$$|u| = 2\mu^{-\frac{1}{2}}(D_{\mathrm{crit}} - D)^{\frac{1}{2}}, \qquad (6.38)$$

which is the standard form of the order parameter for a continuous (second order) phase transition.

In order to check the analytical results, we have performed large scale Langevin simulation of the microscopic system with periodic boundary condition. In two dimensions, the direction of motion in the ordered state $\mathbf{u}/|\mathbf{u}|$ can freely diffuse. Due to always present fluctuations the direction of motion changes in time. Thus, the ordered state is characterized in simulations through the non-vanishing mean speed of the system:

$$\langle|\mathbf{u}|\rangle = \left\langle \sqrt{\left(\frac{1}{N}\sum_{i=1}^{N}\mathbf{v}_i\right)^2} \right\rangle. \qquad (6.39)$$

Here, N is the particle number and $\langle\cdot\rangle$ indicates the temporal average after the system reached the stationary state.

The results of microscopic Langevin simulations in two spatial dimensions $d = 2$ at high particle densities and $L/\varepsilon = 10$ confirm our analysis. They are in a good agreement with the semi-analytical results for the full mean-field system at low μ and with both solutions types at large μ. The temperature components $T_\parallel$ and $T_\perp$ differ significantly at low μ

and the solution of the reduced system is not able to describe the system correctly. But for large μ the differences between the different components become negligible and the different mean-field solution and the numerical results collapse practically on a single line (Fig. 6.5).

In contrast to $d = 1$ no stable disordered solutions at low μ and low noise intensities D were observed (Fig. 6.4). A possible explanation can be the reduced basin of attraction of the disordered solution for $d = 2$ together with the already discussed instability of the disordered solution due to the neglected temperature fluctuations θ_k observed also at $d = 1$.

A more heuristic explanation for the instability of the disordered solution in two spatial dimensions at low μ and low D, is the fact that, in contrast to the one-dimensional case, for $d = 2$ there is no energetic barrier between different directions of motion. In two dimensions the particles can change their direction of motion by continuous angular drift or diffusion. Thus, any finite fluctuation in $\mathbf{u}$ at vanishing noise will be amplified and eventually will lead to perfect alignment of all particle velocities.

Recently, it was shown that the spatially homogeneous state is unstable in systems of interacting self-propelled particles (Bertin et al., 2006; Simha and Ramaswamy, 2002b,a). Thus, the assumption of a spatially homogeneity is only an approximation which is reasonable for sufficiently large ε (upper panel, Fig. 6.6). For $L \gg \varepsilon$ strong density inhomogeneities appear, such as travelling bands (lower panel, Fig. 6.6), which affect the global behavior of the system.

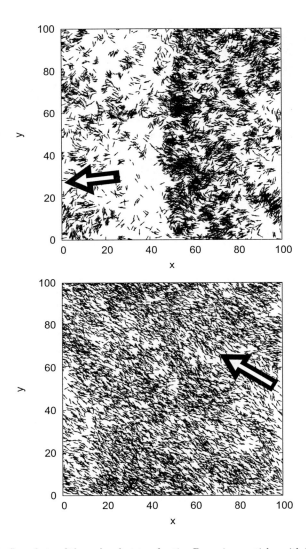

Figure 6.6.: Snapshots of the ordered state of active Brownian particles with RH-friction and velocity alignment for two different values of $\varepsilon = 1.0$ (top) and 5.0 (bottom). The large arrows indicate the mean velocity. (Other parameter values: $N = 8192$, $\mu = 1.0$, $D = 0.1$, $L = 100$)

6.3. Schienbein-Gruler in Two Dimensions

As the last example, we consider a system of active Brownian particles with external noise and a variant of the Schienbein-Gruler friction function in two spatial dimensions ($d = 2$). The friction term used here reads

$$-\gamma(\mathbf{v})\mathbf{v} = -\alpha \left(1 - \frac{v_0}{|\mathbf{v}|}\right) \mathbf{v}. \tag{6.40}$$

Please note that this version of the SG-friction differs from the one introduced in Eq. 3.5. It can be seen as an approximation of (3.5), with the assumption that the heading vector of an active particle is given by the velocity unit vector: $\mathbf{e}_h = \mathbf{e}_v = \mathbf{v}/|\mathbf{v}|$. This approximation does not account for the possibility of backward motion with respect to the heading, because $\mathbf{e}_v$ changes sign if we make the transformation $\mathbf{v} \to -\mathbf{v}$. Thus, for an active particle with the friction given in (6.40) there is no front or back, it always moves to the front.

A consequence of this approximation is an artificial discontinuity and a reflection symmetry of the corresponding velocity potential with respect to $v = 0$: $V(v) = \alpha(|v| - v_0)^2/2$. In fact, the model obtained from inserting (6.40) in (6.2) corresponds to a Schienbein-Gruler speed-model (Chapter 4) with passive fluctuations. Despite the deviations at low velocities, the above friction term conserves the most important feature of the SG-model: the velocity-dependent friction term linear in v.

Following the same approach as in the previous examples we arrive at the following moment equation:

$$\frac{\partial}{\partial t}(v_x^n) = -\frac{\partial}{\partial x}\left(\rho \, \langle v_x^{n+1} \rangle\right) - \frac{\partial}{\partial y}\left(\rho \, \langle v_x^n \, v_y \rangle\right)$$
$$+ n \, \rho \left[\alpha \left(\left\langle \frac{v_x^n}{\sqrt{v_x^2 + v_y^2}} \right\rangle - \langle v_x^n \rangle\right) + \mu(u_{\varepsilon,x} \langle v_x^{n-1} \rangle - \langle v_x^n \rangle)\right]$$
$$+ n \, (n-1) \, D \, \rho \, \langle v_x^{n-2} \rangle. \tag{6.41}$$

The only difference to the corresponding equation for the Rayleigh-Helmholtz friction (6.41) are in the terms associated with the friction function. Due to the absolute value of the velocity $|\mathbf{v}|$ in (6.40), we obtain in the moment evolution equation not only simple linear combination of moments but also the term $\langle v_x^n/\sqrt{v_x^2 + v_y^2} \rangle$, which we have to take care of in order to be able to formulate the mean-field equations. Here, we make the following approximation:

$$\left\langle \frac{v_x^n}{\sqrt{v_x^2 + v_y^2}} \right\rangle \approx \frac{\langle v_x^n \rangle}{\left\langle \sqrt{v_x^2 + v_y^2} \right\rangle} \approx \frac{\langle v_x^n \rangle}{\sqrt{u_x^2 + u_y^2 + T_x + T_y}}. \tag{6.42}$$

For clarity, we will use the following abbreviation for the denominator in the above expression

$$A := A(\mathbf{u}, \mathbf{T}) = \sqrt{u_x^2 + u_y^2 + T_x + T_y}.$$

Thus, we obtain the following system of mean-field equations for the kth-component of the mean-field velocity and temperature $(k = x, y)$:

$$\frac{\partial u_k}{\partial t} + \mathbf{u}\nabla_\mathbf{r} u_k = \left(\frac{\alpha v_0}{A} - \alpha\right) u_k + \mu(u_{\varepsilon,k} - u_k) - \frac{\partial T_k}{\partial x} - \frac{T_k}{\rho}\frac{\partial \rho}{\partial x}, \tag{6.43a}$$

$$\frac{1}{2}\left(\frac{\partial T_k}{\partial t} + \mathbf{u}\nabla_\mathbf{r} T_k\right) = \left(\frac{\alpha v_0}{A} - \alpha - \mu\right) T_k + D - T_k\frac{\partial u_k}{\partial x}, \tag{6.43b}$$

which together with the continuity equation (6.30a) determine the evolution of the mean-field system.

Please note that in contrast to the Rayleigh-Helmholtz friction there was no need for closing the moment equations by neglecting higher moments. This is due to the linear nature of the SG-friction function around the velocity fixed point v_0 and the approximation made in Eq. 6.42.

By considering again the simplest case of a spatially homogeneous system, we arrive at the following set of ordinary differential equations:

$$\frac{du_k}{dt} = \left(\frac{\alpha v_0}{A} - \alpha\right) u_k, \tag{6.44a}$$

$$\frac{1}{2}\frac{dT_k}{dt} = \left(\frac{\alpha v_0}{A} - \alpha - \mu\right) T_k + D. \tag{6.44b}$$

From the symmetry of the temperature equations it follows directly $T_x = T_y$ or by chosing the appropriate reference frame $T_\parallel = T_\perp$. Using the mean velocity squared $u^2 = u_x^2 + u_y^2$ as a variable instead of the individual components (u_x, u_y) the above equations simplify to:

$$\frac{d}{dt}u^2 = 2\alpha\left(\frac{v_0}{A} - 1\right) u^2, \tag{6.45a}$$

$$\frac{1}{2}\frac{dT_k}{dt} = \left(\frac{\alpha v_0}{A} - \alpha - \mu\right) T_k + D. \tag{6.45b}$$

There are two solution of Eq. 6.45a in terms of the temperature components reads:

$$u^2 = 0, \tag{6.46a}$$

$$u^2 = v_0^2 - T_\parallel - T_\perp. \tag{6.46b}$$

The full system of equations has in total three stationary solution: a single ordered state solution with $|u| = \sqrt{u^2} > 0$ and two disordered solutions with $|u| = 0$. The single ordered solution reads:

$$|u|_1 = \sqrt{v_0^2 - \frac{2D}{\mu}}, \qquad T_{k,1} = \frac{D}{\mu}, \tag{6.47}$$

whereas two stationary disordered solutions are given as:

$$|u|_{2,3} = 0, \tag{6.48}$$

$$T_{k,2} = \frac{D}{\alpha + \mu} + \left(\frac{\alpha v_0}{2(\alpha + \mu)}\right)^2 \left[1 + \sqrt{1 + \frac{8D(\alpha + \mu)}{\alpha^2 v_0^2}}\right], \tag{6.49}$$

$$T_{k,3} = \frac{D}{\alpha + \mu} + \left(\frac{\alpha v_0}{2(\alpha + \mu)}\right)^2 \left[1 - \sqrt{1 + \frac{8D(\alpha + \mu)}{\alpha^2 v_0^2}}\right]. \tag{6.50}$$

The second disordered solution $T_{k,3}$ is not considered, as it is always less stable than $T_{k,2}$ and yields unphysical result of vanishing temperature of the disordered phase in the limit $D, \mu \to 0$. For noninteracting self-propelled particles, for which only the disordered state exists, the temperature has to coincide with the average kinetic energy per particle: $v_0^2/2$. It can be easily shown that $T_{k,2}$ satisfies this condition in the respective limit.

For noise intensities below the critical noise intensity

$$D_{\text{crit}} = v_0^2 \mu / 2, \tag{6.51}$$

the homogeneous system is in the ordered state ($|u| = |u|_1$, $T_k = T_{k,1}$), whereas above the critical noise intensity no collective motion takes place ($|u| = |u|_2$, $T_k = T_{k,2}$). For vanishing noise, $D \to 0$, the temperature components T_k vanish and the mean-field speed is given as $|u| = v_0$. For $v_0 = 0$, the ordered state is always unstable and the temperature of the disordered state reduces to $T_k = D/(\alpha + \mu)$. Inserting the critical noise intensity into (6.47) yields for the mean speed

$$|u| = 2^{\frac{1}{2}} \mu^{-\frac{1}{2}} (D - Dc)^{\frac{1}{2}}. \tag{6.52}$$

This corresponds to a second-order phase transition, not only in the limit of large μ, as in the case of Rayleigh-Helmholtz friction.

The full stationary solution of the homogeneous system may be written as:

$$|u| = \begin{cases} |u|_1 & \text{for} \quad D < D_{\text{crit}} \\ 0 & \text{for} \quad D > D_{\text{crit}} \end{cases} \tag{6.53a}$$

$$T_k = \begin{cases} T_{k,1} & \text{for} \quad D < D_{\text{crit}} \\ T_{k,2} & \text{for} \quad D > D_{\text{crit}} \end{cases} \tag{6.53b}$$

As $T_{k,1} = T_{k,2} = v_0^2/2$ for $D = D_{\text{crit}}$ the temperature is continuous but in general not continuously differentiable with respect to the bifurcation parameter (e.g. D, μ).

In order to test the theoretical predictions, we have performed large scale Langevin simulation of the microscopic system either with global coupling $\varepsilon > L$, or with local coupling $\varepsilon = 1/10L$. At high densities, $\tilde{\rho} = (N\varepsilon^2)/L^2 \gg 1$, both situations yield similar results due to the large correlations length in the ordered phase with $l_{\text{corr}} \gg \varepsilon$. Thus, also for local coupling the system can be assumed as spatially homogeneous, if the interaction range ε is not to small with respect to the system size L. All simulations were performed with periodic boundary conditions and with random initial positions and velocities (disordered state).

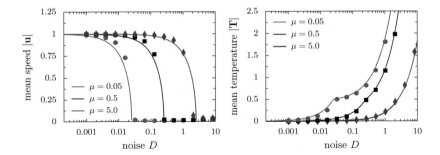

Figure 6.7.: Comparison of the simulation and analytical results for the mean-field speed $|u|$ (left) and temperature T_k (right) versus noise strength D. The results of Langevin simulations are shown as symbols. The solid lines represent the stationary solutions according to Eq. (6.53a). The simulations were performed with periodic boundary condition and with the disordered state as initial condition. Simulation parameters: $\alpha = 1.0$, $v_0 = 1.0$, $\varepsilon = 10.0$, $L = 100.0$, $N = 16384$.

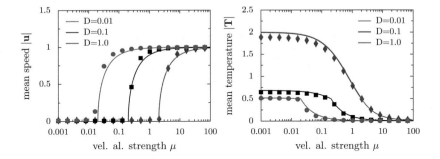

Figure 6.8.: Comparison of the simulation and analytical results for the mean-field speed $|\mathbf{u}|$ (left) and mean temperature $|\mathbf{T}|$ (right) versus velocity alignment μ. The results of Langevin simulations are shown as symbols. The solid lines represent the stationary solutions according to Eq. (6.53a). The simulations were performed with periodic boundary condition and with the disordered state as initial condition. Simulation parameters: $\alpha = 1.0$, $v_0 = 1.0$, $\varepsilon = 10.0$, $L = 100.0$, $N = 16384$.

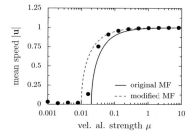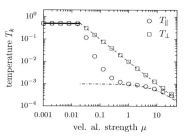

Figure 6.9.: **Left:** Comparison between $|\hat{u}|$ from Eq. (6.55), $|u|$ from Eq. (6.47) and numerical results; **Right:** The corresponding temperature components $\hat{T}_\parallel = D/(\alpha + \mu)$ (dash-dotted line), $\hat{T}_\perp = D/\mu$ (dashed line) and $T_{k,2}$ from (6.49) (solid line) together with results of numerical simulations (symbols). The Langevin simulations were performed with periodic boundary condition and global coupling. Simulation parameters: $\alpha = 10.0$, $v_0 = 1.0$, $N = 16384$.

As shown in Figs. 6.7 and 6.8, the numerical results for the mean speed and the total temperature

$$|\mathbf{T}| = \sqrt{T_\parallel^2 + T_\perp^2} = \sqrt{2}T_k$$

are in good agreement with the theoretical prediction. However, a closer look reveals some systematic deviations. The theoretical results for the temperature T are larger than the ones obtained in the simulations, in particular at low μ. The effect appears for both ordered and disordered state as well as for global and local coupling. This systematic difference vanishes for $v_0 = 0$, thus it may be associated with the moment approximation in the corresponding term of the friction function (Eq. 6.42). Furthermore, as in the case of Rayleigh-Helmholtz friction, the temperature components parallel and perpendicular to the direction of motion are not equal, in contrast to the predictions of the mean-field theory. Whereas the perpendicular component agrees well with the theoretical prediction of $T_\parallel = D/\mu$, we observe in general smaller values of the parallel component.

Starting from the perfectly ordered state in the homogeneous state, it can be easily shown that for small noise intensities the components of the temperature can be approximated as:

$$\hat{T}_\parallel = \frac{D}{\alpha + \mu}, \quad \hat{T}_\perp = \frac{D}{\mu}. \tag{6.54}$$

Using Eq. 6.46b, we obtain then an alternative solution for the mean-field speed:

$$|\hat{u}| = \sqrt{v_0^2 - D\left(\frac{1}{\alpha + \mu} + \frac{1}{\mu}\right)} = \sqrt{v_0^2 - D\left(\frac{\alpha + 2\mu}{\mu(\alpha + \mu)}\right)} \tag{6.55}$$

In the limit of small α or large μ, this result converges towards the original mean-field result. This solution ($|\hat{u}|$, $\hat{T}_\parallel$, $\hat{T}_\perp$) for the ordered state yields a better agreement with numerical results not to close to the critical point but deviates from the numerical solutions close to the onset of collective motion (see Fig. 6.9).

6.4. Discussion of the Mean-Field Approach

In this chapter, we have shown how starting from the microscopic dynamics of active Brownian particles with velocity alignment a corresponding mean-field theory for the coarse grained density $\rho(\mathbf{r}, t)$, velocity $\mathbf{u}(\mathbf{r}, t)$ and temperature $\mathbf{T}(\mathbf{r}, t)$ can systematically be derived.

In general, depending on the details of the active Brownian friction function, approximations have to be applied in order to obtain a self-consistent set of equations. On the one hand, for the simplest possible case of a homogeneous stationary system, it is possible to obtain analytical results, which for a wide range of parameters and different friction functions are in good agreement with numerical results. But on the other hand, the approximations made in the derivation of the mean-field theory, may result for some cases in strong differences between the analytical predictions and numerics.

Nevertheless, the mean-field approach gives us important insights into the behavior of active Brownian particles with velocity alignment. In the case of a (locally) spatially homogeneous system (e.g. global coupling), where the local average velocity which the particles are aligning to equals the mean-field velocity $\mathbf{u}_\varepsilon(\mathbf{r}, t) = \mathbf{u}(\mathbf{r}, t)$, the alignment force term vanishes in the velocity equation. Thus, in this limiting case the velocity alignment acts only on the effective temperature of the active Brownian particle gas. It suppresses the fluctuations around the current mean velocity leading to the stabilization of the ordered state with finite mean-field velocity.

For the homogeneous case, in the limit of large μ where the impact of the individual speed dynamics (friction function) in the temperature becomes negligible, the scaling of the mean speed as order parameter close to the critical noise reads:

$$|u| \sim \left(\frac{D_{\text{crit}} - D}{\mu} \right)^{\frac{1}{2}} . \tag{6.56}$$

Thus, in this limit the onset of collective motion in a spatially homogeneous system takes place via a (continuous) second-order phase transition, irrespective on the details of the velocity dynamics. This result agrees also with previous results obtained for self-propelled particles with constant speed (Peruani et al., 2008).

In general, the nature on the phase transition in the homogeneous system depends on the friction function. For the Rayleigh-Helmholtz friction the mean-field theory predicts a bistability for low μ, which indicates a (discontinuous) first-order phase transition. In one spatial dimension it is easily possible to obtain bistability in numerical simulations due to the presence of the energy barrier. Whereas, in the two-dimensional case, we were not able to obtain the bistability predicted by the mean-field theory. One possible explanation would be an extremely small basin of attraction of the disordered solution in the respective parameter region and the always present finite fluctuations in Langevin simulations (largest system studied: $N = 32768$ with global coupling). Another possible explanation, is the approximation of the probability distribution by a finite number of moments. The consideration of higher orders might lead to changes in the phase space of the system.

For the future, it appears that a more suitable derivation of the mean-field equations should be performed in the "natural" velocity-heading coordinate frame, as introduced in the previous part on individual dynamics. Preliminary results, which go beyond the scope of this work, show that for the general case of fluctuating velocity and direction of motion the resulting equations are quite complex and it might be difficult to obtain analytical results.

The kinetic equations can also be used to analyze the stability of the spatially homogeneous state. An instability of the homogeneous state has been predicted from the analysis of hydrodynamic equations of self-propelled particle systems (Bertin et al., 2006; Simha and Ramaswamy, 2002b,a), which seems to be confirmed by the appearance of strong density fluctuations in our numerical simulations for $\varepsilon \ll L$ (see 6.6). Thus, it should be emphasized that our results hold strictly speaking only in the limit of global coupling and as an approximation also for local coupling, where the interaction range ε is not to small in comparison to the system size L.

7. Collective Motion due to Escape and Pursuit Interactions

7.1. Motivation

A common explanation for the emergence of collective motion in a wide range of animals is that it serves as a protection mechanisms against predators (Krause and Ruxton, 2002). In case of Mormon crickets (*Anabrus simplex*) and desert locusts (*Schistocerca gregaria*), the phenomenon of swarming and collective mass migration is usually associated with high population density and depletion of nutritional resources. The mechanisms driving this migration are still poorly understood. Recent experimental results provide empirical evidence that cannibalism may, surprisingly, facilitate collective motion in mass migrating insects. In field experiments, Simpson et al. (2006) have shown that for Mormon crickets deprived of protein and salt other conspecifics are a major source of these nutrients and cannibalism within such migrating bands is a common event. Thus, the authors suggest that the collective migration of Mormon crickets is not a cooperative phenomenon, but that the insects are rather on a "forced march" driven by other hungry individuals approaching them from the rear.

In controlled experiments with desert locusts, Bazazi et al. (2008) have provided further evidence for cannibalistic interactions driving collective migration. The authors have shown that reducing of individuals' capacity to detect approach of others from behind decreases their probability to start moving, dramatically reduces the mean proportion of moving individuals and significantly increases cannibalism.

Motivated by this intriguing findings, we investigate in this chapter a generic model of individuals with escape and pursuit behavior which may be associated with cannibalism (Romanczuk et al., 2009). We show how these selective repulsion or attraction interactions lead to collective motion of individuals with highly fluctuating speed and analyze the dependence of the model dynamics on the relative strength of individual escape and pursuit response. Directed translational motion, in our model, is a strictly collective (but not cooperative) behavior and may be therefore termed group propulsion. The model offers a novel perspective on possible mechanisms of onset and persistence of collective motion and the resulting migration patterns in nature.

7.2. Model

We model an individual organism as an (active) Brownian particle in two dimensions ($d = 2$). In contrast to ordinary Brownian motion, we assume that an internal energy depot allows the individuals to increase their speed in reaction to external stimuli by conversion of internal energy into energy of motion (see e.g. Ebeling et al., 1999; Erdmann et al., 2000; as well as Chapter 5). For simplicity we assume further on that at all times there is a surplus of internal energy which allows us to neglect the explicit treatment of the energy balance and focus on the spatial dynamics only.

Each individual (particle) obeys the following Langevin dynamics:

$$\dot{\mathbf{r}}_i = \mathbf{v}_i, \quad \dot{\mathbf{v}}_i = -\gamma_0 v_i^{a-1} \mathbf{v}_i + \mathbf{F}_i^s + \sqrt{2D}\boldsymbol{\xi}_i . \tag{7.1}$$

The first term on the left hand side of the velocity equation (7.1) is a friction term with friction coefficient γ_0 and an arbitrary power dependence on velocity represented by $a = 1, 2, 3, \ldots$. The response of individual i to other individuals is described by an effective social force $\mathbf{F}_i^s$. The last term is a non-correlated Gaussian random force with intensity D. A solitary individual ($\mathbf{F}_i^s = 0$) explores its environment by a continuous random walk, where the individual velocity statistics are determined by γ_0, a and D. The parameters are given in arbitrary time and space units T and X: $[\gamma_0] = X^{1-a}T^{a-2}$, $[D] = X^2 T^{-3}$.

Please note that in contrast to previous chapters we do not introduce active motion via a negative friction function into the individual dynamics. In the absence of noise, $D = 0$, the particles cease to move.

The finite-size of individuals is taken into account by fully elastic hardcore collisions with a particle radius $R_{hc} = l_r/2$ (l_r - particle diameter) Brilliantov and Pöschel (2004).

Motivated by the observations on cannibalistic behavior in locusts and crickets, which are in particular vulnerable to attacks from behind (Bazazi et al., 2008), we introduce the following response mechanisms:

- If approached from behind by another individual j, the focal individual i increases its velocity away from it in order to prevent being attacked from behind. We refer to this behavior as *escape* (e).

- If the focal individual "sees" another individual up-front moving away, it increases its velocity in the direction of the escaping individual. We refer to this behavior as *pursuit* (p).

- No response in all other cases.

The response of an individual is determined the following decision algorithm: a) Is there another individual within my sensory range $l_s > l_r$?; b) If yes, is it in front or behind me and c) does it come closer or does it move away? (Fig. 7.1).

Based on the above considerations, we write $\mathbf{F}_i^s$ as a sum of an effective escape and an effective pursuit force:

$$\mathbf{F}_i^s = \mathbf{f}_i^e + \mathbf{f}_i^p. \tag{7.2}$$

The two force terms read

$$\mathbf{f}_i^e = \frac{1}{N_e} \sum_j -\hat{\mathbf{r}}_{ji} K_e(|v_{\mathrm{rel}}|) \Theta_e(v_{\mathrm{rel}}, \mathbf{r}_{ji}), \tag{7.3a}$$

$$\mathbf{f}_i^p = \frac{1}{N_p} \sum_j +\hat{\mathbf{r}}_{ji} K_p(|v_{\mathrm{rel}}|) \Theta_p(v_{\mathrm{rel}}, \mathbf{r}_{ji}) \tag{7.3b}$$

Here, $\hat{\mathbf{r}}_{ji}$ is the unit vector pointing from individual i to individual j:

$$\hat{\mathbf{r}}_{ji} = \frac{\mathbf{r}_{ji}}{r_{ji}} = \frac{\mathbf{r}_j - \mathbf{r}_i}{|\mathbf{r}_j - \mathbf{r}_i|} \tag{7.4}$$

and v_{rel} is the relative velocity of individuals j and i:

$$v_{\text{rel}} = \mathbf{v}_{ji} \cdot \hat{\mathbf{r}}_{ji} = (\mathbf{v}_j - \mathbf{v}_i)\hat{\mathbf{r}}_{ji}. \tag{7.5}$$

The response functions $K_{e,p} \geq 0$ determine the strength of the corresponding interaction. In the following, we assume the response functions proportional to the relative speed:

$$K_{e,p} = \chi_{e,p}|v_{\text{rel}}|, \tag{7.6}$$

with $\chi_{p,e} \geq 0$ being the corresponding interaction strengths. This choice of the response function leads to stronger response to fast approaching/escaping individuals in comparison to those with low relative velocity. It takes into account that, for example, locusts are able to determine accurately relative motion through specialized neuronal structures (Rind and Simmons, 1992; Rind et al., 2008; Rogers et al., 2010).

The $\Theta_{e,p}$-functions can take only two values: either 1 or 0. They are given by a product of Heaviside functions θ (step functions) which define the condition for the escape and pursuit interactions to take place:

$$\Theta_e(r_{ji}, \mathbf{v}_i, v_{\text{rel}}) = \theta(l_s - r_{ji})\theta(r_{ji} - l_r)\theta(-\mathbf{v}_i \cdot \mathbf{r}_{ji})\theta(-v_{\text{rel}}), \tag{7.7}$$

$$\Theta_p(r_{ji}, \mathbf{v}_i, v_{\text{rel}}) = \theta(l_s - r_{ji})\theta(r_{ji} - l_r)\theta(+\mathbf{v}_i \cdot \mathbf{r}_{ji})\theta(+v_{\text{rel}}). \tag{7.8}$$

The product of the first two Heaviside functions is identical for both interaction types. It is 1 only if the individual j is within the social interaction zone defined by the hardcore distance l_r and the sensory range l_s: $l_r < r_{ji} < l_s$. It corresponds to homogenoues response to all individuals within the social interaction zone.

The product of the last two Heaviside functions distinguishes the escape and pursuit interaction. The escape interaction takes place only if individual j is behind the focal individual i ($\mathbf{v}_i \cdot \mathbf{r}_{ji} < 0$) and it is coming closer ($v_{\text{rel}} < 0$). The pursuit interaction takes place only if the individual j is in front of the focal individual i ($\mathbf{v}_i \cdot \mathbf{r}_{ji} > 0$) and it is escaping individual i ($v_{\text{rel}} > 0$).

Finally, both forces in 7.3 are normalized by the respective number of individuals which the i-th individual responses to: the number of individuals which fulfill the escape response conditions

$$N_e = \sum_j \Theta_e(r_{ji}, \mathbf{v}_i, v_{\text{rel}}) \tag{}$$

and the corresponding number for the pursuit response:

$$N_p = \sum_j \Theta_p(r_{ji}, \mathbf{v}_i, v_{\text{rel}}). \tag{}$$

Thus, the individual escape/pursuit response is averaged over the corresponding stimuli. The normalization ensures that the total social forces and the resulting accelerations acting

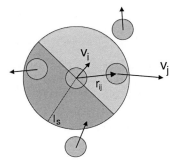

Figure 7.1.: Example of the interaction scheme of individual i. The velocity of all individuals is indicated by vectors. The interaction range of individual i is given by l_s. Green indicates its pursuit interaction range whereas red indicates its escape interaction range. Individual i interacts only with j via the pursuit interaction. It responses with an increase of velocity in the direction of the connecting vector $\mathbf{r}_{ij}/|\mathbf{r}_{ij}|$. All other shown individuals are being ignored as their positions or relative velocities violate the conditions for a pursuit/escape response.

on the individual are bounded to reasonable values even under high density of individuals.

The symmetry of the social interaction is broken in several ways: the interaction acts only on one of the interacting particles (*action≠reaction*); the interactions are direction selective - the particles distinguish between their front ($\mathbf{v}_i \cdot \mathbf{r}_{ji} > 0$) and their back ($\mathbf{v}_i \cdot \mathbf{r}_{ji} < 0$) and between approach ($v_{\mathrm{rel}}$) and escape ($v_{\mathrm{vrel}} > 0$); the strength of interaction to the front and back may be different ($\chi_e \neq \chi_p$). The most important property of the interactions, is their anti-dissipative nature with respect to kinetic energy. Note that $\mathbf{F}_i^s$ leads only to acceleration of individuals and is analogous to the autocatalytic machanism proposed in Bazazi et al. (2008).

7.3. Numerical Results

We will discuss our numerical results in terms of the rescaled density

$$\rho_s = \frac{N l_s^2}{L^2} \,, \tag{7.9}$$

where N is the total particle number, l_s the interaction range and L the size of the simulation domain. All simulation results presented here were obtained with periodic boundary condition.

Numerical simulation reveal that irrespective of the detailed model parameters, the pursuit and escape interactions lead to global collective motion at high particle densities ρ_s and moderate noise intensities D (Fig. 7.2). At low ρ_s however we observe a very

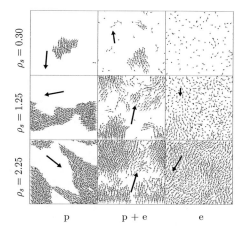

Figure 7.2.: Typical spatial configurations and particle velocities (small arrows) for pursuit
only (p), pursuit+escape (p+e) and escape only (e) cases at different particle
densities $\rho_s = 0.30$, 1.25, 2.25. Mean migration direction and speed U is
indicated by large arrows ($U \approx 0$ for escape only and $\rho_s \ll 1$).

different behavior in dependence on the microscopic details of the model, where the velocity
statistics depend on the relative strength of the escape and pursuit interaction χ_p and χ_e.
For $\chi_e > 0$ and $\chi_p \to 0$ with increasing ρ_s a transition between a disordered state, with
vanishing mean migration speed

$$\langle U \rangle = \frac{1}{N} \left| \sum_i \mathbf{v}_i \right| = 0 \qquad (7.10)$$

and an ordered state with $\langle U \rangle > 0$ takes place. This resembles similar transitions reported
for self-propelled particles with velocity alignment (Fig. 7.2, 7.3) Vicsek et al. (1995). With
increasing χ_p the transition shifts to lower ρ_s until it vanishes. For $\chi_p > 0$ and $\chi_e \to 0$
there is no dependence of $\langle U \rangle$ on ρ_s.

In order to explain the above findings we have to distinguish the influence of escape and
pursuit interactions independently, for that purpose we continue with an analysis of the
extreme cases: $\chi_p = 0$, $\chi_e > 0$ (only escape) and $\chi_e = 0$, $\chi_p > 0$ (only pursuit).

Escape Only

In the escape only case the particles try to keep their distance with respect to individuals
approaching from behind. To the front only interactions via the short range repulsion
take place. At low ρ_s after an escape response the probability of interaction within the
characteristic time of velocity relaxation vanishes and the particles are able to reorient

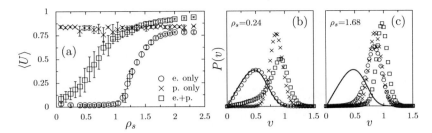

Figure 7.3.: (a) Mean velocity $\langle U \rangle$ for escape-only ($\bigcirc$) $\chi_e=10$, $\chi_p=0$, pursuit-only ($\times$) $\chi_e=0$, $\chi_p=10$ and symmetric escape+pursuit ($\square$) $\chi_e=\chi_p=10$ vs. ρ_s obtained from numerical simulations with periodic boundary conditions ($\gamma_0=1$, $D=0.05$, $a=3$, $l_r=2$, $l_s=4$; only translational solutions were considered; errorbars represent one std. deviation). Particle speed distribution $P(v)$ for the different interaction types in comparison with the analytical solution for non-interacting Brownian particles (solid line) at $\rho_s=0.24$, 1.68 (b,c).

themselves (disordered state). As ρ_s increases the frequency of escape interactions increases and the particles are able to correlate their velocities on several interaction length scales but subensembles may move in different directions. We observe a transition to the ordered state via an active fluid like state ($\rho_s \approx 1.25$). In the ordered state all particles are able to correlate their direction of motion. At all ρ_s the density remains almost spatially homogeneous.

Pursuit Only

In the case of pursuit-only interaction the dynamics change dramatically. Already at low ρ_s we observe a highly inhomogeneous state, initiated by formation of small compact particle clusters performing coherent translational motion. As there is no escape interaction the density of the clusters is only limited by the hardcore radius. At moderate noise intensities the clusters are highly stable and a process of cluster fusion can be observed where larger clusters absorb smaller clusters and solitary particles. For most parameter values, the dominant stationary configuration in a finite system with periodic boundary condition and moderate noise, is a single large cluster performing translational motion (Fig. 7.2). The migration speed $\langle U \rangle$ in Fig. 7.3 is given by the mean speed of a single cluster $\langle u \rangle = |\sum_{i \in \text{cluster}} \mathbf{v}_i|/N_{\text{cluster}}$. For large clusters $\langle u \rangle$ becomes independent of the cluster size and therefore independent of ρ_s (7.31). An intriguing feature of the pursuit-only interaction is the spontaneous formation of large scale vortices out of random initial conditions (Fig. 7.4), due to collisions of translational clusters moving in opposite direction. The emergence of vortex structures in our model is in particular remarkable because so far they have only been reported for systems of self-propelled particles with confinement, or attracting potential, respectively. Czirok et al. (1996); D'Orsogna et al. (2006); Vollmer et al. (2006). Here the pursuit behavior acts in a sense as both: a propulsion mechanism and an asymmetric attraction. This pursuit-only vortices may be highly

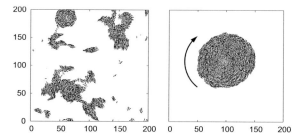

Figure 7.4.: Vortex formation for the pursuit only ($\chi_e = 0$, $\chi_p = 10$) out of random initial distribution with periodic boundary condition ($\rho_s = 0.72$, $D = 0.05$, $\gamma_0 = 1$). Individual particle velocities are shown by vectors. **left:** At $t = 100$ a vortex structure nucleated at $x \approx 50$, $y \approx 180$ due to a collision of clusters moving in opposite direction. **right:** Finally at $t = 2400$ a single vortex structure emerges through absorption of the last remaining moving cluster. The large arrow indicates the rotation direction of the vortex.

stable with lifetimes of an individual rotating cluster exceeding 10^4 time units. In fact for certain parameter settings they appear as the dominant structures in the pursuit-only case.

Based on preliminary numerical results suggesting an increasing stability of the vortex motion with increasing size of the vortex, we performed a systematic numerical analysis of the stability of the rotational motion for different number of particles N in a vortex. For each N we initialized 100 vortices by appropriate initial conditions and measured their rotational order parameter given by the averaged normalized angular momentum:

$$\bar{L} = \frac{1}{N} \sum_{i=1}^{N} \frac{(\mathbf{r}_i - \mathbf{r}_v) \times (\mathbf{v}_i - \mathbf{u}_v)}{|(\mathbf{r}_i - \mathbf{r}_v||\mathbf{v}_i - \mathbf{u}_v|}, \tag{7.11}$$

with $\mathbf{r}_v = 1/N \sum_N \mathbf{r}_i$ being the center of mass of a rotating cluster and $\mathbf{u}_v = 1/N \sum_N \mathbf{v}_i$ being the center of mass velocity. The rotational order parameter is 1 for perfect rotational motion of all particles and 0 for pure translational motion. The decay of the vortex was defined by a critical threshold value for the rotational order parameter: $\bar{L}_c = 0.3$. Typically we observed a breakdown of rotational motion with a transition to translational motion of the cluster as soon as $\bar{L}$ decreased below ≈ 0.5. Thus, the choice of $\bar{L}_c$ is arbitrary, but different values of $\bar{L}_c$ do not change the qualitative results.

We summarized the results of the numerical simulations in Fig. 7.5. The number of vortices N_v, which survived for at least 200 time units, shows a steep initial increase with increasing vortex size N and saturates quickly at a fraction close to 100% (for the simulation parameters used $N_v/N \approx 0.9$, Fig. 7.5a). The average lifetime $\langle \tau_v \rangle$ of the N_v vortices increases initially as well but after passing a maximum starts to decreases with further increasing N (Fig. 7.5b). Based on this results we conclude that there exists an optimal size of rotational clusters. The large fluctuations in the numerical results for the

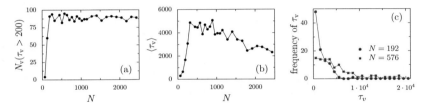

Figure 7.5.: Analysis of vortex stability for the pursuit only case with respect to the vortex size N (number of particles). (a) Number of vortices N_v with a lifetime exceeding $\tau_v = 200$ from a total of 100 initiated vortices. (b) Average lifetime $\langle \tau_v \rangle$ of a vortex obtained from all vortices which survived for at least 200 time units. (c) Vortex lifetime τ_v histograms for two different sizes: $N = 192$ and $N = 576$. Simulation parameters: $\chi_p = 4.0$, $D = 0.2$, $l_s = 4.0$, $R_{hc} = 1.0$.

average survival time may be associated with the low ensemble size and the exponential-like distribution of individual lifetimes (Fig. 7.5c), where the standard deviation is in the same order of magnitude as the mean $\langle \tau_v \rangle$.

The existence of an optimal vortex size can be explained based on following reasoning: The rotational center of a vortex can be regarded as a topological defect. It is not stationary but moves randomly within the rotational cluster due to fluctuations in positions and velocities of individual particles constituting the cluster. For small clusters, the rotational center has a high probability to reach the boundary of the rotational cluster. This leads to a complete breakdown of rotational motion and to the onset of translational motion of the cluster. With increasing size (area) of the vortex this "escape" of the rotational center to the boundary becomes less probable which explains the initial increase in vortex stability. On the other hand, assuming a constant angular velocity of the vortex an increasing size (increasing radius) of a rotational cluster would lead to increasing tangential velocities of particles at the boundary. This in turn is likely to destabilize the vortex motion and increases the probability of vortex breakdown, induced by a splitting up of outer particle layers. Although, this explanation seems reasonable only a further detailed analysis of the vortex motion, which goes beyond the scope of this work, can provide the final answer.

The analysis of the dynamics shows that both interactions — escape and pursuit — lead to active motion of groups but have an opposite impact on the density distribution. Whereas escape leads in general to a homogenization of density within the system, pursuit facilitates the formation of density inhomogeneities. This leads us to the insight that the actual escape+pursuit dynamics where $\chi_p, \chi_e > 0$ is a competition of the two opposite effects with respect on the impact on the particle density. The stability of moving clusters in this simple model is determined by the relative ratio of the interaction strengths. In general for the escape+pursuit case at low ρ_s we observe fast formation of actively moving particle clusters with complex behavior: fusion and break up of clusters due to cluster collisions as well as spontaneous break up of clusters due to fluctuations.

7.4. Pair Approximation

In order to determine the velocity of individual clusters, we consider the smallest cluster which shows directed translational motion: a particle pair $(1, 2)$. We assume particle 1 being in the front of particle 2 and $|\mathbf{r}_{12}| < l_s$ at all times. The transformation of Eq. 7.1 into polar coordinates with $\mathbf{v}_i = (v_i \cos \delta\varphi_i, v_i \sin \delta\varphi_i)$, where $\delta\varphi_i$ is defined as the angle between $\mathbf{v}_i$ and $\hat{\mathbf{r}}_{12}$, yields:

$$\frac{d}{dt} v_1 = -\gamma_0 v_1^a + \chi_e |v_{12}| \theta(-v_{12}) \cos \delta\varphi_1 + \sqrt{2D}\xi_{v,1} \tag{7.12a}$$

$$\frac{d}{dt} v_2 = -\gamma_0 v_2^a + \chi_p |v_{12}| \theta(+v_{12}) \cos \delta\varphi_2 + \sqrt{2D}\xi_{v,2} \tag{7.12b}$$

$$\frac{d}{dt} \delta\varphi_1 = \frac{1}{v_1} \left(-\chi_e |v_{12}| \theta(-v_{12}) \sin \delta\varphi_1 + \sqrt{2D}\xi_{\varphi,1} \right) \tag{7.12c}$$

$$\frac{d}{dt} \delta\varphi_2 = \frac{1}{v_2} \left(-\chi_p |v_{12}| \theta(+v_{12}) \sin \delta\varphi_2 + \sqrt{2D}\xi_{\varphi,2} \right). \tag{7.12d}$$

Here, $v_{12} = v_1 - v_2$ is the relative velocity of the two particles and $\xi_{\varphi,i}, \xi_{v,i}$ represents the transformed noise variables. For $-\pi/2 < \varphi_1, \varphi_2 < \pi/2$ the escape and pursuit interaction leads to an increase of either v_1 or v_2 in order to harmonize the speed of the slower particle with the faster one. In addition the interaction stabilizes the translational motion along $\hat{\mathbf{r}}_{12}$, i.e. $\langle \delta\varphi_i \rangle \to 0$ (Eqs. 7.12). After the system relaxes to a stationary state ($\hat{\mathbf{r}}_{12}$ varies slowly in time), we end up with almost one-dimensional translational motion of the particle pair with slowly diffusing direction of motion defined by the direction of $\hat{\mathbf{r}}_{12}$. Please note that for elastic hardcore interaction the total energy and momentum of the particle pair does not change during collisions.

In order to obtain equations of motion for the particle pair in the stationary translational state, we assume that the particle velocities are given by the mean velocity u of the particle pair plus a small deviation: $v_i = u + \delta v_i$. Furthermore the particles are assumed to have approximately the same heading so that $\delta\varphi_i \ll 1$. Performing a Taylor expansion of right hand sides of Eqs. (7.12) and keeping only terms up to the order δv_i, $\delta\varphi_i$ yields

$$\frac{d}{dt} (u + \delta v_1) = -\gamma_0 (u^a + au^{a-1}\delta v_1) + \chi_e |v_{12}| \theta(-v_{12}) + \sqrt{2D}\xi_{v,1}, \tag{7.13a}$$

$$\frac{d}{dt} (u + \delta v_2) = -\gamma_0 (u^a + au^{a-1}\delta v_2) + \chi_p |v_{12}| \theta(+v_{12}) + \sqrt{2D}\xi_{v,2}, \tag{7.13b}$$

$$\frac{d}{dt} \delta\varphi_1 = \left(\frac{1}{u} - \frac{1}{u^2}\delta v_1 \right) \left(-\chi_e |v_{12}| \theta(-v_{12})\delta\varphi_1 + \sqrt{2D}\xi_{\varphi,1} \right), \tag{7.13c}$$

$$\frac{d}{dt} \delta\varphi_2 = \left(\frac{1}{u} - \frac{1}{u^2}\delta v_2 \right) \left(-\chi_p |v_{12}| \theta(+v_{12})\delta\varphi_2 + \sqrt{2D}\xi_{\varphi,2} \right), \tag{7.13d}$$

with $v_{12} = \delta v_1 - \delta v_2$.

From the addition of the two velocity equations (Eqs (7.13a) and (7.13b)) we obtain

$$\frac{d}{dt} \left(u + \frac{\delta v_1 + \delta v_2}{2} \right) = -\gamma_0 (u^a - au^{a-1}\frac{\delta v_1 + \delta v_2}{2})$$
$$+ \frac{\chi_e}{2} |v_{12}| \theta(-v_{12}) + \frac{\chi_p}{2} |v_{12}| \theta(+v_{12}) + \sqrt{2D}(\xi_{v,1} + \xi_{v,2}). \tag{7.14}$$

Time averaging assuming of the vanishing mean of the speed deviations $\langle \delta v_i \rangle = 0$ yields:

$$\frac{d}{dt}u = -\gamma_0 u^a + \frac{\chi_e}{2}\langle |v_{12}|\theta(-v_{12})\rangle + \frac{\chi_p}{2}\langle |v_{12}|\theta(+v_{12})\rangle. \tag{7.15}$$

In the symmetric case, where $\chi_e = \chi_p = \chi$, the social force terms on the right hand side can be summed up and we obtain:

$$\frac{d}{dt}u = -\gamma_0 u^a + \frac{\chi}{2}\langle |v_{12}|\rangle . \tag{7.16}$$

There is permanently a social force acting on one of the particles. In order to evaluate the expectation value of the social force we assume that at all times one of the particles moves with the mean velocity u ($\delta v_{1,2} = 0$) whereas the velocity of the second particle deviates by δv as a result of the non-correlated stochastic force which yields $|v_{12}| = |\delta v|$. We approximate the expectation value of the relative speed $|\langle |v_{12}|\rangle$, by considering the speed deviations as discrete increments taken from a Gaussian distribution with zero mean and variance $\sigma_1^2 = 2D\tau$ (Wiener process). Here τ is the relaxation time of the escape+pursuit interaction $\tau = \chi^{-1}$.

$$\langle |\delta v|\rangle_{1d} = \int_{-\infty}^{\infty} \frac{|\xi|}{\sqrt{4\pi D\chi^{-1}}}e^{\frac{-\chi\xi^2}{4D}}d\xi = 2\sqrt{\frac{D}{\pi\chi}} . \tag{7.17}$$

Alternatively the same result for $|\delta v|$ can be obtained by considering $|\delta v|$ as a stochastic variable under the action of a linear restoring force $-\chi\delta v$ subject to white Gaussian fluctuations.

Thus, in the limit of small angular deviations $\delta\varphi_i \approx 0$ (quasi one-dimensional motion) the evolution of the averaged pair velocity for symmetric escape and pursuit reads:

$$\frac{d}{dt}u = -\gamma_0 u^a + \sqrt{\frac{\chi D}{\pi}}, \tag{7.18}$$

and for the stationary pair velocity u_{ep}^s we obtain:

$$u_{ep}^s = \left(\frac{1}{\gamma_0}\sqrt{\frac{\chi D}{\pi}}\right)^{\frac{1}{a}} . \tag{7.19}$$

A detailed analysis of the pair dynamics reveals that in two dimensions the particle distance increases slowly for the symmetric escape and pursuit case due to fluctuations, which are not fully compensated by the social force (see Fig. 7.6). This slow increase in $|\mathbf{r}_{12}|$ actually stabilizes the quasi one dimensional motion as it decreases $\delta\varphi_i$ but leads finally to a breakdown of collective motion as soon as the particle distance increases above the sensory range l_s.

Interestingly, in the case of sufficiently large l_s, where the particle sense each other for a long time despite the (slow) increase in $|\mathbf{r}_{12}|$, the mean square displacement of the particle pair $\langle \mathbf{r}_p^2 \rangle$ shows a behavior which is neither purely diffusive nor purely ballistic over extremely long time scales. This is due to the competition of angular fluctuations and the slowly increasing particle distance which counteracts deviations in the direction

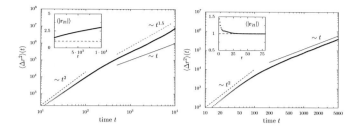

Figure 7.6.: Mean squared displacement of particle pairs versus time for symmetric escape and pursuit (**left**) and for pursuit only (**right**). The insets show the mean distance between the two particles where the dashed line indicates the minimal possible distance given by the particle diameter l_r.

of motion of the particle pair due to individual fluctuations.

Particle pairs with a non-increasing particle distance are only possible for pursuit dominated dynamics, in particular for the pursuit only case: $\chi_p > 0$ and $\chi_e = 0$. In general a breakup of a cluster can occur either because the inter-particle distance becomes larger than the interaction range or due angular fluctuations, which violate the condition for the social interaction: $|\delta\varphi| < \pi/2$. In principle this is also possible for the pursuit only case, especially for low χ_p and large D, but with increasing values of χ_p the pair break-up becomes an extremely rare event.

Due to the attracting nature of the pursuit interaction, for sufficiently large χ_p, the inter-particle distance tends to decrease for $|\mathbf{r}_{12}| > l_r$. Thus, the limiting particle distance is given by the particle diameter l_r. In a stationary case a particle pair with pursuit only behaves as a single active particle and the mean squared displacement exhibits the expected behavior of normal diffusion for $t \to \infty$ (see Fig. 7.6).

The time averaged evolution of the mean velocity for $\chi_e = 0$ (Eq. (7.15)) reads:

$$\frac{d}{dt}u = -\gamma_0 u^a + \frac{\chi_p}{2}\langle|v_{12}|\theta(+v_{12})\rangle . \tag{7.20}$$

The pursuit force acts only on the pursuing particle at times where it is slower than the leading (first) particle. In the quasi one-dimensional situation, when the second (pursuing) particle is moving faster than the first one, it will eventually collide with it and a momentum exchange takes place which immediately makes the first particle faster than the second one. For point particles ($l_r = 0$) the second (faster) particle will overtake the first one, which corresponds for identical particles similarly to an exchange in momentum. Thus, it is evident that the velocity deviations will be larger in the pursuit only case. In order to evaluate the expectation value of the relative velocity we consider now both velocity deviations δv_1 and δv_2 as stochastic variables. In analogy to the escape and pursuit case, we approximate the expectation value of the relative speed $\langle|v_{12}|\rangle$ by considering it as discrete increments taken from a Gaussian distribution with zero mean and variance $\sigma_2^2 = 2\sigma_1^2 = 4D\tau_p$, with $\tau_p = \chi_p^{-1}$. The factor 2 follows from the assumption that $\langle|v_{12}|\rangle$ is

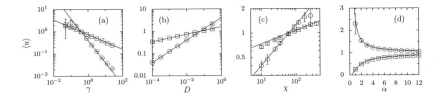

Figure 7.7.: Pair velocities $\langle u \rangle$ for symmetric escape and pursuit. Comparison of numerical results for $a = 1$ ($\bigcirc$) and $a = 3$ ($\square$) with the result of Eq. 7.19 (solid lines): (a) $\langle u \rangle$ over friction coefficient γ_0, (b) noise intensity D, (c) and interaction strength χ. (d) $\langle u \rangle$ vs. friction function exponent a. Here we distinguish two cases $A > 1$ ($\bigcirc$) and $A < 1$ ($\square$), with $A = \sqrt{D\chi/\pi}/\gamma_0$.

given by the difference of two independent stochastic processes with variance σ_1^2.

The resulting evolution equation for the velocity in the pursuit only case reads:

$$\frac{d}{dt}u = -\gamma_0 u^a + \sqrt{\frac{\chi_p D}{2\pi}}, \qquad (7.21)$$

and the stationary velocity is given as

$$u_p^s = \left(\frac{1}{\gamma_0} \sqrt{\frac{\chi_p D}{2\pi}} \right)^{\frac{1}{a}}. \qquad (7.22)$$

A comparison of the analytical results for the averaged pair velocity confirms our analytical findings for a large range of parameter values for the symmetric escape and pursuit (Fig. 7.7) and the pursuit only case (Fig. 7.8). The numerical results were obtained for particle pairs which were stable during the entire simulations time. The stationary mean speed was obtained by a temporal average of 50 independent runs after the pair velocity became stationary. Deviations from the analytical result become apperent in parameter regions where the assumption on quasi one-dimensional motion of the particle pair does not hold as for example for large noise strengths D or low pair velocities.

7.5. Mean Field Approach

In this section a mean-field analysis of our model system for the symmetric escape and pursuit interaction ($\chi_e = \chi_p = \chi$) will be performed.

The starting point for the analysis is the evolution of the velocity probability density of an individual particle embedded in a gas of other identical particles (see Chapter 6). For simplicity, we assume directly a spatially homogeneous system, which is a reasonable approximation of the system state at high densities ρ_s, or within a large cluster of particles (Fig. 7.2). The evolution of the velocity probability density function (PDF) $p(\mathbf{v}, t)$ is given

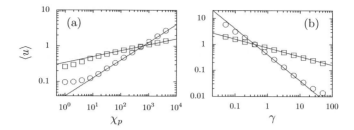

Figure 7.8.: Average pair velocities $\langle u \rangle$ for the pursuit only case. Comparison of numerical results for $a = 1$ ($\bigcirc$) and $a = 3$ ($\square$) with the result of Eq. 7.22 (solid lines): (a) $\langle u \rangle$ versus pursuit strength χ_p, (b) friction coefficient γ_0.

by

$$\partial_t p(\mathbf{v}, t) = -\nabla_{\mathbf{v}}(-\gamma_0 v^{a-1} \mathbf{v} p + \tilde{\mathbf{F}}^s p) + D \Delta_{\mathbf{r}} p. \tag{7.23}$$

With the assumption of only pairwise interactions, the mean-field force can be written as

$$\tilde{\mathbf{F}}^s(\mathbf{v}) = \int d\mathbf{v}' \mathbf{F}^s(\mathbf{v}, \mathbf{v}') p(\mathbf{v}'). \tag{7.24}$$

The combination of Eq. 7.23 and 7.24 represents an nonlinear Fokker-Planck equation which is, in general, not solvable exactly, but it is possible to obtain approximate solutions by formulation of evolution equations for the moments of the probability density p. Since the system is, in general, far from equilibrium we expect p to be non-Gaussian. In principle, for the correct description we have to take infinitely many moments (Pawula, 1967). Here we make an approximation and truncate the expansion by considering only moments up to the second order ($n \leq 2$).

In the following only the equations for the x-component will be given, the corresponding equations for the y-component can be obtained by interchanging x and y. The moment equations obtained from Eq. 7.23 read

$$\partial_t \langle v_x^n \rangle = -n\gamma_0 \left(\langle v_x^{n-1+a} \rangle + \langle v_x^n v_y^{a-1} \rangle \right) \\ + n \langle v_x^{n-1} \tilde{F}_x^s \rangle + n(n+1) D \langle v^{n-2} \rangle, \tag{7.25}$$

with the moment definitions according to Eqs. 6.3, 6.5, 6.6. The components of mean velocity $\mathbf{u} = (u_x, u_y)$ are given by the first moments

$$\langle v_x \rangle = \int v_x P(\mathbf{v}, t) d\mathbf{v} = u_x. \tag{7.26}$$

The absolute value of the mean-field velocity $u = |\mathbf{u}|$ in a homogeneoues system corresponds to the mean migration speed $\langle U \rangle$ (7.10) measured in computer simulations.

The individual velocity of a particle can be written in terms of $\mathbf{u}$ and individual velocity fluctuations around the mean $\delta\mathbf{v}$.

$$v_x = u_x + \delta v_x. \tag{7.27}$$

Assuming a symmetric distribution of velocity fluctuations around the mean velocity and independence of fluctuations in x and y we obtain: $\langle\delta v_x\rangle = \langle\delta v_y\rangle = \langle\delta v_x\delta v_y\rangle = 0$ and $T_i = \langle\delta v_i^2\rangle$. T_i gives us a mean-field measure of the velocity fluctuations. In addition, we use for the expectation value of the social force $\bar{F}_x^s = \langle\tilde{F}_x^s\rangle$.

The Eqs. 7.25 simplify to a system of first order differential equations for the evolution of U_x and T_x, which for the linear friction $a = 1$ read

$$\partial_t u_x = -\gamma_0 u_x + \bar{F}_x^s, \tag{7.28a}$$

$$\frac{1}{2}\partial_t T_x = -\gamma_0 T_x + D, \tag{7.28b}$$

and for $a = 3$ become

$$\partial_t u_x = -\gamma_0 u_x(u_x^2 + u_y^2 + 3T_x + T_y) + \bar{F}_x^s, \tag{7.29a}$$

$$\frac{1}{2}\partial_t T_x = -\gamma_0 T_x(3u_x^2 + u_y^2 + T_x + T_y) + D. \tag{7.29b}$$

For $a = 3$ we neglected higher order fluctuations in T_i (see Chapter 6).

In order to make the equations Eqs. 7.28 and 7.29 self-consistent an expression for $\mathbf{F}^s = (\bar{F}_x^s, \bar{F}_y^s)$ has to be derived.

We approximate the solution of Eq. 7.24 in the high-density limit by considering a single particle with a fluctuating velocity given in Eq. 7.27 embedded in a homogeneous particle bath which moves with constant mean velocity $\mathbf{u} = (u_x, 0)$. The PDF of surrounding particle velocities in Eq. 7.24 is given by a δ-function: $p(\mathbf{v'}) = \delta(\mathbf{v'} - \mathbf{u})$. Averaging over the interaction range for symmetric pursuit and escape interaction yields:

$$\bar{F}^s = 2\chi \int_{-\infty}^{\infty} \delta v_{2d}Q(\delta v_{2d})\theta(-\delta v)d\delta v_{2d}. \tag{7.30}$$

Here, $Q(\delta v_{2d})$ is the probability density of the speed fluctuations of the focal particle, which we assume, as in the pair approximation, to be a Gaussian with variance $\sigma_{2d}^2 = 2\sigma_{1d}^2 = 4D\chi^{-1}$ (Eq. 7.17). The Heaviside function $\theta(-\delta v_{2d})$ takes into account the acceleration of the focal particle only if the surrounding particle bath moves faster than the focal particle. Integrating (7.30) gives

$$\bar{F}^s = 2\sqrt{2\frac{\chi D}{\pi}}. \tag{7.31}$$

Please note that $\bar{F}^s$ is independent of the number of the interaction partners N_e, N_p due to the normalization in Eq. 7.3. The additional factor $\sqrt{2}$ with respect to the particle pair approximation in the symmetric escape and pursuit case results from the consideration of velocity fluctuations in two dimensions. The resulting mean-field force vector is given by $\bar{\mathbf{F}}^s = (\bar{F}^s\cos\vartheta, \bar{F}^s\sin\vartheta)$ where ϑ is the difference angle between the orientation of the focal particle and the mean-field velocity vector (stationary case: $\vartheta \to 0$).

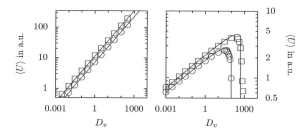

Figure 7.9.: Mean migration speed $\langle U \rangle$ vs noise intensity D. Comparison of the mean-field solutions (Eqs. 7.28, 7.29; solid lines) with numerical simulations of the full system at $\rho_s = 66, 7$ for $\chi = 20$ ($\bigcirc$) and $\chi = 50$ ($\square$). **left:** linear case with $a = 1$, **right:** nonlinear case with $a = 3$.

So far, we considered the focal particle aligned with the mean velocity **u** except for small fluctuations $\delta\mathbf{v}$. This is reasonable at small noise intensities, but with increasing D the rotational diffusion of individual particles may lead to a loss of orientational order, which in turn decreases the cross-section of the escape and pursuit interaction. As the magnitude of the mean velocity u is a direct measure of the orientation we introduce an effective cross-section $C(u)$

$$\bar{F}^s = 2\sqrt{2\frac{\chi D}{\pi}}C(u). \tag{7.32}$$

We obtain $C(u)$ in Eq. 7.32 by calculating the probability of individual turning angles $|\varphi| < \pi/2$ by free rotational diffusion within the characteristic interaction time $\tau = \chi^{-1}$.

$$C(u) = 2 \int\limits_{0}^{\pi/2} \frac{1}{2\pi D_\varphi \chi^{-1}} \exp\left(-\frac{\chi\varphi^2}{2D_\varphi}\right) d\varphi \tag{7.33}$$

$$= \mathrm{erf}\left(\frac{\pi u}{4}\sqrt{\frac{\chi}{D}}\right), \tag{7.34}$$

with $D_\varphi = D/u^2$.

In order to obtain the correct probability, we considered φ on the interval $-\infty, \infty$. This choice ensures an outflux of probability out of the interval $-\pi/2, \pi/2$ for large D (small u), which corresponds to a loss of orientational information due to rotational diffusion ($C(u) \to 0$). For small D and large u the turning angles φ are strongly confined around zero: the effective cross section $C(u)$ is close to unity.

By inserting Eq. 7.32 into Eqs. 7.28 and 7.29, we obtain a self-consistent systems of equations for the linear and nonlinear case. Unfortunately, it is not possible to obtain the stationary solutions as analytical expressions in contrast to the velocity-alignment models studied in Chapter 6. Here, we used standard root-finding procedures to find the stationary solution, whereby a single solution branch was obtained by taking reasonable initial conditions (Fig. 7.9).

The results of the mean-field theory confirm the scaling behavior of the particle pair approximation with respect to χ, γ_0, a and D at low noise intensities D. At large D however the mean-field theory predicts a qualitatively different behavior of the mean speed u for the linear ($a = 1$) and the nonlinear ($a = 3$) case. For linear friction u increases monotonously with D, whereas for the nonlinear friction there exists a maximum of u. Above the critical noise the collective migration breaks down which results in a fast decrease of u. Numerical simulations of the full system for large ρ_s in agreement with the mean-field assumption confirm the results of the mean-field theory in qualitative and quantitative terms as shown in Fig. 7.9.

The absence of a breakdown of collective motion with increasing noise strength D for linear friction can be understood from the fact that the squared mean speed of collective motion scales linearly with the noise strength: $u^2 \sim D$. As a consequence, the effective rotational diffusion does not depend on D: $D_\varphi = D/u^2 = const.$. For the nonlinear case the squared mean speed scales as: $u^2 \sim D^{1/a}$. Thus, the rotational diffusion increases with the noise strength: $D_\varphi = D/u^2 \sim D^{1-1/a}$, leading to a breakdown of collective motion at finite D.

7.6. Selective Attraction and Repulsion of Self-Propelled Particles

In this section, we discuss briefly a generalization of the escape and pursuit model for active particles which move with a fixed speed and can change only their direction of motion (heading). Such particles are usually referred to as self-propelled particles.

Although we restrict ourselves here only the case of $v_i = v_0 = const.$, the general conclusions of this section apply also to the active Brownian particle models studied in Chapters 3, 4 and 6 at low velocity fluctuation strengths.

We consider a system of N self-propelled particles in two spatial dimensions, which move with a constant velocity (speed) $v_0 > 0$. The evolution of the system is determined by the following set of stochastic differential equations:

$$\dot{\mathbf{r}}_i = v_0 \mathbf{e}_{h,i}(t) = v_0 \begin{pmatrix} \cos\varphi_i(t) \\ \sin\varphi_i(t) \end{pmatrix}, \tag{7.35}$$

$$\dot{\varphi}_i = \frac{1}{v_0}\left(F_{i,\varphi} + \sqrt{2D_\varphi}\xi_\varphi \right), \tag{7.36}$$

with φ_i being the (polar) angle, which determines the heading of the focal particle. The temporal evolution of φ_i is determined by the turning of the individual due to social interactions $F_{i,\varphi}$ and random (angular) fluctuations with the intensity D_φ.

The (angular) social force is given by the projection of the total social force vector $F_{i,\varphi} = \mathbf{F}_i \mathbf{e}_{\varphi,i}$ on the angular degree of freedom with $\mathbf{e}_\varphi = (-\sin\varphi_i, \cos\varphi_i)$. The noise variable ξ_φ is given by Gaussian white noise with zero mean and vanishing temporal correlations.

For active Brownian particles (or self-propelled) particles with a stationary velocity $v_0 > 0$ and vanishing velocity fluctuations the decay of $\mathbf{vv}$-correlations is much slower then for Brownian particles at comparable strengths of the stochastic forces. The persistent motion of individual particles with constant speed favors strongly the emergence of ordered states with broken symmetry due to interactions (Ramaswamy, 2010). For example, inelastic collisions in Brownian particle systems (e.g. granular media; Brilliantov and Pöschel, 2004) may lead to density instabilities, but never to large scale collective motion. This changes if we consider self-propelled particles interacting via inelastic collisions as shown recently by Grossman et al. (2008). The dissipation of the kinetic energy associated with the relative velocity of the colliding particles results in an effective alignment of the self-propelled particles and, as a consequence, in large scale collective motion.

In our case, the consideration of self-propelled particles allows us to drop the spatial inhomogeneity in the escape and pursuit interactions: The response of the focal individual i to individual j depends only on the sign of the relative velocity $v_{rel} = \mathbf{v}_{ji}\hat{\mathbf{r}}_{ji}$ and not – in contrast to the original Brownian model (Sec. 7.2) – on the relative spatial position of individual j.

The total social force is given by a sum of three components in analogy to the escape/pursuit interaction and the short-ranged hardcore interaction from the previous section:

$$\mathbf{F}_i = \mathbf{f}_m + \mathbf{f}_a + \mathbf{f}_r. \tag{7.37}$$

The first two forces read:

$$\mathbf{f}_m = \frac{1}{N_+} \sum_{j=1}^{N} \mu_a |v_{rel}| \hat{\mathbf{r}}_{ji} \theta(l_s - r_{ji}) \theta(r_{ji} - R_{hc}) \theta(+v_{rel}), \tag{7.38}$$

$$\mathbf{f}_a = \frac{1}{N_-} \sum_{j=1}^{N} \mu_m |v_{rel}| \hat{\mathbf{r}}_{ji} \theta(l_s - r_{ji}) \theta(r_{ji} - R_{hc}) \theta(-v_{rel}). \tag{7.39}$$

The terms $\mu_{m,a}|v_{rel}|$ determine the turning rates due to the respective interaction. Please note that for $\mu_a \leq 0$ and $\mu_m \geq 0$ the response terms are similar to those of the original model as defined in Eq. (7.6). In contrast to the original model, we do not formulate any restrictions on the sign of the response strengths $\mu_{m,a}$ ($\mu_{m,a} \in \mathbb{R}$), which leads to different dynamical regimes, discussed further below.

The last force is the short range repulsion defined as:

$$\mathbf{f}_r = -\frac{\mu_r}{N_r} \sum_{j=1}^{N} \hat{\mathbf{r}}_{ji} \theta(R_{hc} - |r_{ji}|), \tag{7.40}$$

where $\mu_r \geq 0$ is a constant repulsive turning rate. We restrict here to the cases where the repulsion turning rate μ_r is larger then the maximal turning rate for the other two interaction types:

$$\mu_r > \text{Max} \left(\mu_m |v_{rel}|, \mu_a |v_{rel}| \right) = 2v_0 \text{Max}(\mu_m, \mu_a) \tag{7.41}$$

All social interactions are normalized by the respective number of individuals for the corresponding interaction type: N_a, N_m and N_r.

The social forces $\mathbf{f}_{a/m}$ lead to either repulsion (attraction) to approaching individuals for $\mu_a < 0$ ($\mu_a > 0$) or repulsion (attraction) to individuals moving away $\mu_m < 0$ ($\mu_m > 0$).

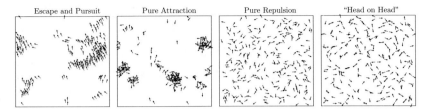

Figure 7.10.: Examples of spatial configurations for different regimes: "Escape and Pursuit" ($\mu_a = -1.0$, $\mu_m = 1.0$), "Pure Attraction" ($\mu_a = 1.0$, $\mu_m = 1.0$), "Pure Repulsion" ($\mu_a = -1.0$, $\mu_m = -1.0$) and "Head on Head" ($\mu_a = +1.0$, $\mu_m = -1.0$).

In the $\mu_m \mu_a$-parameter space we distinguish four large regions corresponding to different behavior types:

i. **Pure Repulsion:** repulsion from approaching and moving away individuals: $\mu_a < 0$ and $\mu_m < 0$.

ii. **Escape and Pursuit:** repulsion from approaching individuals $\mu_a < 0$, attraction to moving away individuals $\mu_m > 0$.

iii. **"Head on Head":** attraction to approaching individuals $\mu_a > 0$, repulsion from moving away individuals $\mu_m < 0$.

iv. **Pure Attraction:** attraction to approaching and moving away individuals: $\mu_a > 0$ and $\mu_m > 0$.

There exist also the special cases with $\mu_{m/a} = 0$ and $\mu_{a/m} > 0$ ($\mu_{a/m} < 0$), which correspond to a selective attraction (repulsion) only to approaching/moving away individuals.

We refer to the situation $\mu_a < 0$ and $\mu_m > 0$ as "Escape and Pursuit", due similar behavior as in the original Brownian particle model. For $\mu_a > 0$ and $\mu_m < 0$ the social forces lead to a preference to move towards other individuals which already are coming closer and therefore favor (in particular at low densities) frontal collisions between individuals. We refer to this regime as "Head on Head".

In order to characterize the behavior of the above model in the $\mu_m \mu_a$-plane, we have performed systematic numerical simulations of the system for varying interactions strengths $-1 \leq \mu_{a,m} \leq 1$ and different densities ρ. The density is determined by the size of the simulation domain with periodic boundary conditions L and the total number of individuals N.

We measure the degree of collective motion in the system by the time averaged velocity order parameter in the stationary state $\langle \Phi \rangle$ defined as the normalized mean speed:

$$\langle \Phi \rangle = \left\langle \frac{1}{N v_0} \left| \sum_{i=1}^{N} \mathbf{v}_i \right| \right\rangle \, , \tag{7.42}$$

where $\langle \cdot \rangle$ denotes the temporal average. In addition, we measure the spatial inhomogeneity (clustering) by the time averaged scaled neighbor number

$$\langle n \rangle = \left\langle \frac{\bar{n} - n_{\min}}{n_{\max} - n_{\min}} \right\rangle, \tag{7.43}$$

with $\bar{n}$ being the average number of neighbours within the sensory range of an individual at a given time step. The density dependent numbers $n_{\min}$ and $n_{\max}$ define the minimal and maximal expectation values for the measured neighbor numbers corresponding to a random homogeneous distribution and to closest packing of individuals (discs) with a diameter l_r:

$$n_{\min} = \max \left(\pi \rho_s - 1, \ 0 \right), \qquad n_{\max} = \eta_{2d} \frac{4 l_s^2}{l_r^2}. \tag{7.44}$$

Here, ρ_s is the rescaled density defined in Eq. 7.9 and $\eta_{2d} = \pi/(2\sqrt{3}) \approx 0.907$ is the packing fraction for the closest packing of discs in two spatial dimensions. The term -1 in the definition of $n_{\min}$ takes into account that the focal particle is not being counted as its own neighbor.

In Fig. 7.11 the results of numerical simulations for the order parameter and the neighbor number are shown in dependence on μ_m and μ_a. Large scale collective motion emerges at sufficiently high densities essentially only for $\mu_a < 0$ (repulsion from approaching individuals) and for $\mu_m > 0$ (attraction to moving away individuals). This parameter region corresponds to the original escape-pursuit scheme. This is also reflected in the neighbor number $\langle n \rangle$ in this region, which indicates strong clustering for large μ_m (pursuit dominated dynamics). Interestingly, the neighbor number in the "Escape+Pursuit" regime, seems to follow a non-monotonous dependence on μ_a with a maximum at a finite $\mu_a < 0$.

We observe also strong clustering for $\mu_{a/m} > 0$ (attraction to everyone) without any collective motion. In this region the particles form quasi-stationary clusters (swarms) with random motion of individuals within and slow diffusive motion of the of individual cluster center of mass.

In the regime $\mu_{a/m} < 0$ (repulsion from everyone), a low degree of collective motion is observed only for $|\mu_m| \ll |\mu_a|$. Whereas in "Head-on-Head" regime no collective motion was observed.

This brief discussion should be considered only as a starting point for future analysis of the generalized model. It offers an interesting alternative to other models because with the two different social forces acting on an individual ($\mathbf{f}_a$ and $\mathbf{f}_m$) and without any explicit velocity alignment, we may obtain very different dynamical structures, such as stationary, disordered clusters, moving clusters or almost spatially homogeneous collective motion of particles reminiscent of an active incompressible fluid.

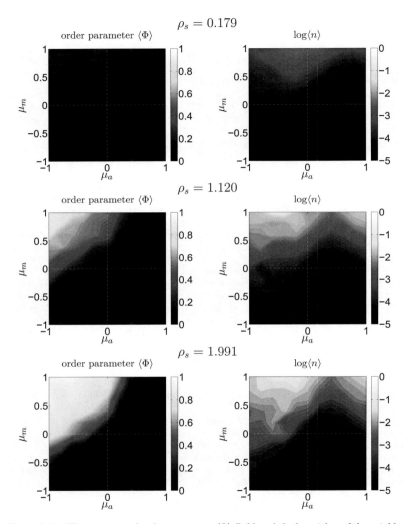

Figure 7.11.: The mean speed order parameter $\langle\Phi\rangle$ (left) and the logarithm of the neighbor order parameter $\log\langle n\rangle$ (right) versus μ_m and μ_a for different densities ρ_s. The dashed lines indicate the zero axes.

8. Collective Motion of Locusts Swarms

In this chapter, we discuss modelling of recent experiments on the impact of the nutritional state on collective dynamics of desert locusts by a modified escape and pursuit model. In the first section, we give a brief description of the performed experiment, the obtained data and the main experimental results. In the second section, we extend the escape and pursuit model, discussed in Chapter 7, in order to account for the observed speed dynamics of individual locusts. After formulating different possible model variants, we proceed with the discussion of the parameter choice based on experimental data. In the end, we go beyond a pure comparison of our modelling results with the experimental data, by making predictions on the impact of the nutritional condition on the onset of collective motion in locusts swarms.

8.1. Experimental Set-Up and Results

The experiments were performed by Sapideh Bazazi at the Department of Zoology, Oxford University under the supervision of Iain D. Couzin (EEB, Princeton University) and Graham K. Taylor (Zoology Department, University of Oxford).

Healthy intact freshly moulted gregarious desert locusts (*Schistocerca gregaria*) were reared under controlled condition on three different diets: low-protein, balanced and high-protein for 48h. After 48h either individual locusts ("single locust experiment") or a group of 15 insects ("group experiment") were placed in a ring-shaped experimental arena with 80cm diameter, walls 52.5cm high and a central dome with 17.5cm diameter (as described in Buhl et al., 2006). The motion of the locusts was then filmed for 8h using a digital video camera (Canon XM2). For each diet and experiment type (single/group) thirty trials were carried out, resulting in a total of 180 trials (Bazazi et al., 2010).

Automated digital tracking software, which captured images at a rate of 5 per second ($\Delta t = 0.2s$), was used to analyze the video material and extract position data of individual locusts. The tracking of individual insects in the single locust experiments provides continuous position data for the entire duration of the experiment. Examples of trajectory samples obtained from single locusts experiments are shown in Fig. 8.1. In group experiments it was not possible to distinguish individual insects at all times due to tracking errors and mixing up of close-by individuals by the tracking software. Therefore, only quantities averaged over the entire group can be extracted from the group experiment data.

From the position data it is possible to obtain the individual displacement speed by a simple difference ansatz

$$s(t_i) = \frac{|\Delta \mathbf{r}(t_i)|}{\Delta t}$$

with

$$|\Delta \mathbf{r}|(t_i) = \sqrt{(x(t_i) - x(t_{i-1}))^2 + (y(t_i) - y(t_{i-1}))^2}$$

being the displacement within the time step $\Delta t = t_i - t_{i-1}$.

The analysis of the data reveals large differences in the motion behavior of individuals, even if reared under the same nutritional condition. A general feature of all speed distributions obtained from single locusts experiments is a large peak close at $s = 0$. The spatial tracking data does not provide us with any information on the heading of individuals. Thus the velocity (speed) data corresponds to sampling of the velocity distribution in the laboratory frame of reference (Cartesian coordinates). We have shown that the Cartesian velocity distributions of particles with active fluctuations may exhibit a sharp peak at the origin. In principle this effect may be the cause the increase of the speed distribution for $s \to 0$. From direct observation of the locusts behavior during experiments a stop and go behavior can be observed[1], where marching events are interrupted by extended stationary phases. A preliminary analysis of the data and corresponding theoretical model suggests that the actual stop and go dynamics as well as a possible tracking errors have a much stronger effect and represent the dominating mechanism responsible for the peak at $s = 0$.

The uncertainties in the automated tracking make it difficult to distinguish stationary (not moving individuals) from slowly moving individuals. Thus, based on the data, we define in the following moving individuals as all individuals with a measured speed larger than 1.5cm/s (dashed line in Fig. 8.2).

Many locusts trajectories show a strong interaction with arena walls, which affects the measured speed data. In order to minimize the impact of the arena boundary we have applied an additional filtering of the original data, by considering only trajectory samples with a minimal distance of Δl_w to the walls. The filtering of wall interactions leads for many data samples to the appearance of a distinct local maximum at finite speeds $s > 0$, which we associate with the preferred marching speed of freely moving locusts. An increase of Δl_w beyond 3cm does not change the shape of the speed distributions and the positions of the maxima but result in much smaller data samples leading to lower quality of the filtered distributions. Thus we used $\Delta l_w = 3$cm throughout the analysis of the experimental data in order to filter out the wall interactions.

The Role of Diet in Single and Group Experiments

The impact of different diets on the motion of locusts was examined through the mean speed of moving individuals and the proportion of time spent moving by individuals. In single locusts experiments no significant effect of the diet on the mean speed (Fig. 8.3) as well as on the proportion of time spent moving was found (Bazazi et al., 2010). For all diets the mean speed of individuals as well as the proportion of time spent moving increased slowly over the full duration of the experiment. Leading to significant difference between the begin (first two hours) and end (last two hours) of the experiment (Fig. 8.3).

In contrast to the single locust experiment, the results from group experiments reveal a significant impact of the diet on the mean speed of individuals. Individuals in groups fed on a low protein diet move faster than those on a high protein diet. A balanced diet shows speed levels intermediate to low- and high-protein diets (Fig. 8.4). The diet has also a significant impact on the proportion of time spent moving at the begin of the experiment.

---

[1]S. Bazazi, personal communication

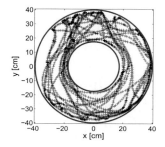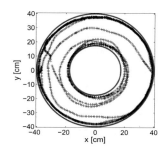

Figure 8.1.: Trajectory samples obtained from the single locust experiment. The two plots show position recordings (black crosses) of two different individuals over a time interval of 10mins at 3h after the begin of the experiment. The red (gray) circles indicate the arena walls.

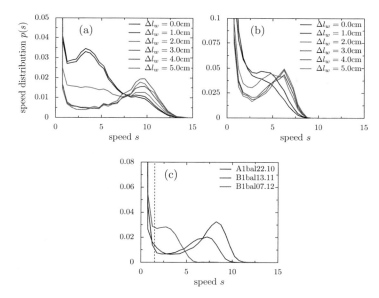

Figure 8.2.: Individual speed distributions from single locusts experiments (balanced protein diet). Effect of filtering away data points within a distance Δl_w to the wall for two different individuals: A1bal16.2 (a) and A1bal07.11 (b). Comparison of speed distribution obtained from three different individuals after filtering out the wall interactions ($\Delta l_w = 3$cm). The dashed line indicates $s = 1.5$cm/s.

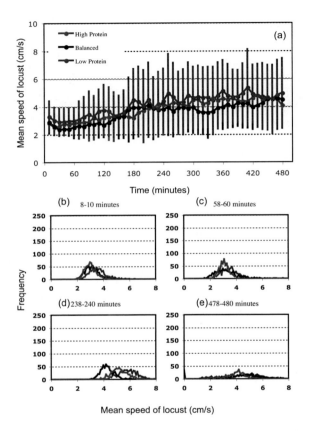

Figure 8.3.: The mean speed of a locust when alone. (a) The time series showing the mean speed (cm/s) of a locust when alone in the arena for each of the diet treatments: high protein (red), balanced (black) and low protein (blue). Thirty experimental trials were carried out for each treatment. Error bars show one s.d. (b-e) The frequency distributions for the mean speed (cm/s) of a locust fed on each diet calculated over different two-minute time-windows during the experiment: (b) 8-10 min, (c) 58-60 min, (d) 238-240 min and (e) 478-480 min, illustrating how the speeds of individuals fed on different diets change over time. (Figure reprinted from Bazazi et al., 2010)

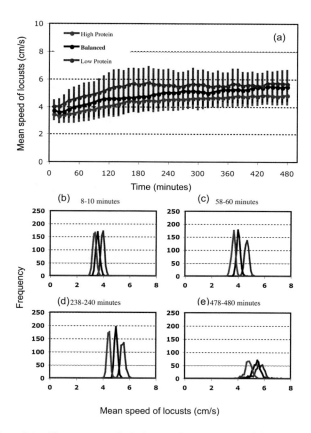

Figure 8.4.: The mean speed of a locust when in a group. (a) The time series showing the mean speed (cm/s) of a locust when in a group of individuals fed on different diet treatments: high protein (red), balanced (black) and low protein (blue). Thirty experimental trials were carried out for each treatment. Error bars show ± one s.d. (b-e) The frequency distributions for the mean speed (cm/s) of a locust in a group fed on different diets calculated over two-minute time-windows at different points during the experiment: (b) 8-10 min, (c) 58-60 min, (d) 238-240 min and (e) 478-480 min, illustrating how the speeds of individuals fed on different diets change over time. (Figure reprinted from Bazazi et al., 2010)

At the end of the experiment the mean proportion of moving locusts for all diets reaches an asymptotic value between 0.5-0.7.

The surprising difference in the impact of the nutritional state on the motion dynamics of solitary individuals versus individuals in a group suggests that the diet affects mainly the social interaction between insects. In the original Brownian particle model we have shown that the escape and pursuit interactions lead effectively to an acceleration of individuals which results in increasing average speed with increasing $\chi_{e,p}$ (Eq. 7.19, Fig. 7.7). Based on this two observations we suggest that the nutritional state is correlated with the strength of social interaction, such that larger protein deprivation (lower protein content in the diet) can be associated with larger $\chi_{e,p}$. In order to test this hypothesis we will develop in the next section a locusts model based on the original escape and pursuit model and the novel experimental data.

8.2. Modelling of Locusts Swarms

Each individual is described as a point particle, which can move actively and interact with neighbours within a finite sensory range l_s. The response of individual i to other individuals is described by an effective social force $\mathbf{F}_i^s$. The total social force is a sum of three different components:

$$\mathbf{F}_i^s = \mathbf{f}_i^e + \mathbf{f}_i^p + \mathbf{f}_i^r.$$

The first two terms describe the empirically motivated escape and pursuit interactions with spatially inhomogeneous response introduced in Section 7.2 (Eqs. 7.3, 7.6 and 7.7). The third force term, $\mathbf{f}_i^r$, is a short range repulsion as introduced Section 7.6 (Eq. 7.40) which accounts for the preferred minimal distance to other individuals. Each force is normalized by the respective number of interactions ($N_{e,p,r}$).

Please note that the conditions for the escape and pursuit interactions (Eq. 7.7) ensure that there are no escape/pursuit interactions with stationary individuals $v_j = 0$ (Buhl et al., 2006).

We account for the observed "stop and go" behavior of individuals by introducing two different kinematic states of individuals with stochastic transition rates between them: a resting (or not-moving) state with $v_i = 0$ (N-state) and an active (or moving) state with $v_i > 0$ (M-state).

A moving individual in the M-state is described by the following set of Langevin equations for its speed s_i and its polar angle (direction of motion) φ_i:

$$\dot{s}_i = f_i(s_i) + F_{v,i} + \sqrt{2D_s}\xi_{s,i}, \tag{8.1}$$

$$\dot{\varphi}_i = \frac{1}{s_i}\left(F_{\varphi,i} + \sqrt{2D_\varphi}\xi_{\varphi,i}\right), \tag{8.2}$$

with $f_i(s_i)$ being an individual velocity dependent propulsion function. For our model, we have chosen the Rayleigh-Helmholtz function (Ch. 4):

$$f_i(s_i) = \alpha_i s_i - \beta s_i^3. \tag{8.3}$$

A comparison of the Rayleigh-Helmholtz function and the corresponding speed distributions with individual locusts data suggests that it yields a better fit to experimental data than, for example, the (linear) Schienbein-Gruler function.

The components of the social force acting on the speed and the angle are determined by the projections of the social force vector on the unit vector in the direction of motion $\mathbf{e}_{v,i} = (\cos\varphi_i, \sin\varphi_i)$ and the unit vector in the angular direction $\mathbf{e}_{\varphi,i} = (-\sin\varphi_i, \cos\varphi_i)$:

$$F_{v,i}^s = \mathbf{F}_i^s \cdot \mathbf{e}_{v,i} \qquad F_{\varphi,i}^s = \mathbf{F}_i^s \cdot \mathbf{e}_{\varphi,i}. \tag{8.4}$$

The turning rate $\dot{\varphi}$ is inversely proportional to the speed (see Eq. (3.7)). Furthermore, we assume that the speed and orientation of individuals change due to independent random processes corresponding to active noise discussed in Section 3.1.2. As the speed can only be positive $s_i > 0$, we impose a reflecting boundary condition at $s_i = 0$. D_s and D_φ are the corresponding speed and orientation noise intensities and ξ is a normally distributed random number with $\langle \xi(t) \rangle = 0$ and $\langle \xi(i,t)\xi(j,t') \rangle = \delta_{ij}\delta(t - t')$.

The dynamics of a solitary individual ($\mathbf{F}_i^s = 0$) correspond directly to the speed model with Rayleigh-Helmholtz friction function discussed in Chapter 4. The prefered speed $s_{0,i}$ is a fixed point of the speed dynamics and in the absence of perturbations (e.g. noise or social interactions) each individual will move with its own preferred speed: $s_{0,i} = \sqrt{\alpha_i/\beta_i}$. For finite noise strengths ($D_s > 0$) the speed fluctuates randomly, but the maximum of the speed probability distribution of an i-th individual is located at $s_{0,i}$.

In the N-state the individuals are at rest ($s_i = 0$) and keep their last orientation in the M-state $\varphi_i = const$. For a solitary individual $\mathbf{F}_i^s = 0$ the new direction of motion after a NM-transition is chosen randomly, whereas, for socially interacting individual it is chosen to be the direction of the total social force $\mathbf{F}_i^s/|\mathbf{F}_i^s|$ acting on the individual at the time of the transition. The initial speed after a NM-transition is set to $s = s_{0,i}$.

8.2.1. Transition Rates between Kinematic States

The change of the kinematic state for a solitary individual is influenced by various internal and external factors, which are effectively described as stochastic processes with transition rates κ_{nm} and κ_{mn} (probabilities of transition per unit time). For an individual within a group, both transition rates may depend on its social interactions with other individuals in its vicinity. In particular, in the context of our work they might depend on the nutritional condition either directly or through social interactions. The proportion of moving individuals obtained from experimental data shows a weak transient dependence on the diet. It approaches towards the end of the experiment an asymptotic value of $0.5 - 0.7$ independent of the diet ($t > 300min$). Furthermore, previous experiments show no significant dependence of the fraction of moving individuals on the density Buhl et al. (2006). In the simplest model variant studied here, we assumed both rates as constant: $\kappa_{mn} = const.$ and $\kappa_{nm} = const.$. In order to confirm that this simplification does not have a major impact on the model predictions for the onset of collective motion, we have tested different extensions of the model with transition rates dependent on the diet, or on the strength of the social force acting on an individual. All tested variants yield similar qualitative and quantitative predictions on the onset of collective motion.

Constant rates

For constant transition rates ($\kappa_{mn} = const.$, $\kappa_{nm} = const.$) the temporal evolution of the probabilities to find an individual in the moving state q_m and in the resting state q_n obey the following set of coupled differential equations:

$$\frac{dq_m}{dt} = -\kappa_{mn}q_m + \kappa_{nm}q_n \tag{8.5}$$

$$\frac{dq_n}{dt} = \kappa_{mn}q_m - \kappa_{nm}q_n \ , \tag{8.6}$$

with $q_m + q_n = 1$.

By setting the temporal derivatives in Eqs. 8.5,8.6 to zero, the stationary values of the probabilities can be easily calculated to

$$q_m = \frac{1}{1 + \kappa_{mn}/\kappa_{nm}}, \qquad q_n = \frac{1}{1 + \kappa_{nm}/\kappa_{mn}} = 1 - q_m \ . \tag{8.7}$$

This result holds only in the case where resting individuals with vanishing speed ($s = 0$) can be clearly distinguish from moving individuals ($s > 0$) which is not possible in the experiments due to the tracking errors. A moving individual in experiments was defined as an individual with a measured speed $s > 1.5cm/s$.

The fraction of moving (and not moving) individuals in simulations was measured according to the definition used in the experiments. For given rates κ_{mn} and κ_{nm}, the measured probabilities differ slightly from the theoretical result in Eqs. 8.7. This was taken into account for the choice of the transition rates in order to get the best possible agreement between experiments and simulations.

In this simple model variant, the (theoretical) fraction of moving individuals is constant and independent on the nutritional state and the social interactions. But the proportion of moving individuals measured according to the experimental definition (see previous section) shows a weak increase with the interaction strength χ in simulation (Fig. 8.6). This is because the average speeds of individuals increase with χ, which leads to a weak decrease of the probability of slowly moving individuals with $s < 1.5cm/s$.

A visualization of the speed time series from the two state model with constant transition rates in comparison with an experimental individual time series is shown in Fig. 8.5.

Social force dependent rates

A more complicated variant of the model assumes a dependence of the individual transition rates on the total strength of the social force acting on the respective individual $F^s = |\mathbf{F}^s|_i$. We used the following ansatz for the social force dependent rates:

$$\kappa_{mn}(F^s) = \kappa_{mn}^0 \exp(-\nu F^s), \qquad \kappa_{nm}(F^s) = \kappa_{nm}^0 \exp(\nu F^s). \tag{8.8}$$

For a solitary (not socially interacting) individual $F^s = 0$ the transition rates are given by the constants κ_{mn}^0 and κ_{nm}^0. The impact of the social interactions on the transition rates is determined by an additional constant $\nu \geq 0$. With this modification a socially interacting resting individual is more likely to start moving. Whereas, a socially interacting moving individual is less likely to stop. The mathematical formulation of the dependence on F^s

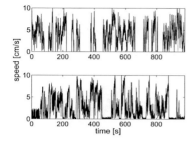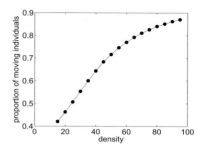

Figure 8.5.: **Left:** Example of individual speed time series obtained from simulation of the model (top; constant rates model) in comparison with a typical experimental time series (bottom). **Right:** Simulation results on proportion of moving individuals versus density for social force dependent rates.

ensures $\kappa_{mn}, \kappa_{nm} > 0$ for arbitrary values of F^s and is motivated by the dependence of transitions rates in stochastic bistable systems on an additional (small) bias. In our context the social force acts as an bias towards the moving state.

As F^s varies in time, as well as between individuals, we can define an average force strength $\langle F^s \rangle$ given by an average over all individuals and over time. This average force, and the corresponding average transition rates, will depend on the density of individuals ρ and on the interaction strength χ.

The assumption that the probability of being in the active state depends on social interactions appears biologically reasonable, but seems to disagree with previous experimental results. Buhl et al. (2006) report no significant dependence of the proportion of moving locusts on the density, which contrasts with the strong dependence observed in the simulations of this model variant shown in Fig. 8.5.

Transition rates directly dependent on the nutritional state

Another possible model variant can be obtained by considering different constant transition rates for different nutritional states. In this model variant the fraction of moving individuals depends directly on the nutritional state but not on the social interactions, and thus, not on the density of individuals. It is easily possible to choose values of κ_{nm} and κ_{mn} for the different diets in such way that a good agreement with the experimental results can be achieved. However, we have to rule out this variant as it does not agree with the experimental results. The fraction of moving individuals in the group experiments corresponds directly to the proportion of time spent moving in the single locust experiment. Thus, if the transition rates depend only on the nutritional state, we should observe for the different diets a similar difference in the time spent moving by individual locusts as the proportion of moving individuals in the group experiment, which is not the case (Bazazi et al., 2010).

Despite this disagreement, we will discuss it for the sake of completeness. In particular, we will show that it yields similar predictions on the impact of the nutritional state on the onset of collective motion.

8.2.2. Parameter Choice Based on Experimental Results

After introducing the model, we proceed with a discussion of the parameter choice. The large number of model parameters makes it virtually impossible to study the full parameter space. But as we are interested in the modelling of the particular experiments presented in Sec. 8.1, we choose the model parameter based on the experimental locusts data and on previous studies.

The noise intensities were set to $D_v = 2cm^2/s^3$ and $D_\varphi = 2cm^2/s^3$, these values correspond approximately to the speed and angle fluctuations estimated from analysis of the individual data from experiments $\sim 1 - 10cm^3/s$, and $\sim 1 - 5cm^3/s^2$, respectively.

The range of repulsion was set to the typical locusts size $l_r = 4cm$, whereas the sensory range of interaction was set to $l_s = 14cm$ according to Buhl et al. (2006). In order to ensure a strong short range repulsion, we set $\chi_r = 10cm/s^2$.

The group experiments were performed at a density $\rho = 37m^{-1}$, thus for comparison we set the density of particles in our simulations to this value.

Furthermore, in agreement with previous experimental results Bazazi et al. (2008) we consider escape dominated social response in our simulations, with a constant ratio $\chi_e/\chi_p = 5$.

In order to account for the large differences in individual speed dynamics, we take random preferred speeds in the active state from an uniform distribution

$$s_{0,i} \in [1.5cm/s, 8cm/s].$$

By setting $\alpha_i = 0.14$ (mean value of α obtained from fits of Rayleigh friction to individual experiments), we obtain the second parameter in the Rayleigh friction (Eq. 8.3) as $\beta_i = \alpha_i/s_{0,i}^2$. In order to ensure that the result do not depend on a particular ensemble configuration of $s_{0,i}$, all presented results were obtained from at least 10 independent simulation runs.

Constant transition rates

For the simplest variant of the model with constant rates, the absolute values of κ_{mn}, κ_{nm} were chosen to correspond approximately to the inverse average times spent in both states observed in experiments. Via a fine tuning of the relative values, we ensure that the (stationary) proportion of moving individuals agrees approximately with the average value observed in experiments for the three different diets (Fig. 8.6b). The used values were: $\kappa_{mn} = 0.025$ s^{-1} and $\kappa_{nm} = 0.040$ s^{-1}.

Thus, in the constant rate model we end up with a single free parameter χ, which we assumed to be correlated with the nutritional state. We performed systematic simulations for increasing values of χ. From comparison of the average speed and average proportion of moving individuals we identified three different χ values, which yield a good quantitative agreement with the respective diets: $\chi_{\text{high}} = 4.0$ for high protein diet, $\chi_{\text{bal}} = 8.0$ for balanced protein and $\chi_{\text{low}} = 12.0$ for the low protein diet (Fig. 8.6a).

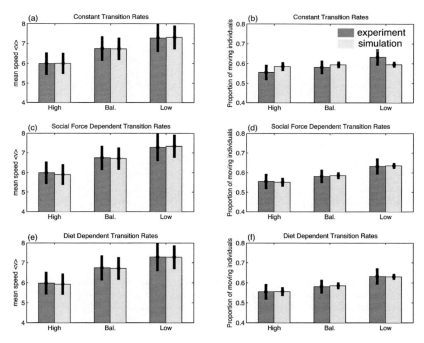

Figure 8.6.: Comparison of the mean speeds of moving individuals (left panels) and the proportion of moving individuals (right panels) between the experiments and the model with constant transition rates (a-b), transition rates dependent on the social interaction strength (c-d), and with diet dependent transition rates (e-f). Kinematic state transition parameters for the model with constant transition rates are $\kappa_{mn} = 0.025$, $\kappa_{nm} = 0.04$; for the model with transition rates dependent on the social interaction: $\nu = 0.015$, $\kappa_{mn}^{0} = 0.019$, $\kappa_{nm}^{0} = 0.015$; for the model with diet dependent transition rates: the MN-transition rates for the different diets are $\kappa_{mn}^{(high)} = 0.014$, $\kappa_{mn}^{(bal)} = 0.013$, $\kappa_{mn}^{(low)} = 0.011$. The NM-transition rate for all diets: $\kappa_{nm} = 0.02$. All other parameters as discussed in Sec. 8.2.2.

Social force dependent transition rates

In the social force dependent variant, we have to choose the additional parameter ν which accounts for the impact of the social force on the two transition rates. This parameter was chosen in combination with χ-values so that both the proportion of moving individuals as well as the mean speed show a good agreement with the data from group experiments (Fig 8.6c,d). The estimated parameter values are: $\nu = 0.015$, $\kappa_{nm}^0 = 0.019$, $\kappa_{mn}^0 = 0.015$.

Diet dependent transition rates

For the diet dependent transition rate model, we first adjusted the (constant) transition rates for each diet in order to match the proportion of moving individuals with the experimental results (Fig. 8.6f). Using this values, we performed systematic simulations for the different transition rates/diets. In this way, we were able to obtain the corresponding χ-values by matching the mean speeds of individuals (Fig. 8.6e). The used transition rate values for the different diets are: $\kappa_{mn}^{(high)} = 0.014$, $\kappa_{mn}^{(bal)} = 0.013$ and $\kappa_{mn}^{(low)} = 0.011$. The NM-transition rate for all diets is $\kappa_{nm} = 0.020$. The estimated interaction strengths associated with the diet are: $\chi_{high} = 3.9$, $\chi_{bal} = 7.8$ and $\chi_{low} = 10.0$.

8.3. Model Results — Speed Distributions and Onset of Collective Motion

As a validation of our model and the corresponding parameter choice, we compared the time averaged speed distribution obtained from group experiments with corresponding results of our model.

Here, we show only the speed distributions for the simplest model variant with constant transition, diet-independent rates (Fig. 8.7). The two different model variants, yield the same qualitative results. Whereby, it should be noted that in general they lead to growing of the second maximum at finite speed, due to the increasing fraction of moving individuals with increasing χ (increasing protein deprivation).

The experimental ensemble speed distributions show, in analogy to the individual speed distributions, a large peak at $s = 0$ corresponding to stationary individuals and an additional maximum at finite speeds $s > 0$. The position of this second maximum depends on the nutritional condition of the locusts: it shifts with increased protein deprivation to larger speeds. This diet-dependent shift is responsible for the corresponding increase in the mean speed $\langle s \rangle$ of moving individuals.

A comparison of the ensemble speed distributions from experiments and simulations shows the same qualitative shape with a large peak at $s = 0$ and a second maximum at $s \approx 5 - 10 \text{cm/s}$ (Fig. 8.7). Furthermore, an increase in the interaction strength χ leads to a shift of the second maximum, observed in experiments with increasing protein deprivation. The quantitative differences between the simulation and experimental distributions, in particular the different peak heights of the maximum, are due to the deviations at small speeds ($s = 0.5 - 2 \text{cm/s}$). The discontinuous shape of the distributions obtained from simulations originates from the discrete formulation of the model (kinematic states), whereas the smoothness of the experimental curves can be due to errors (from noisy position recordings) in measuring the speed of individuals within the ensemble.

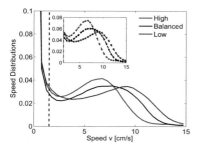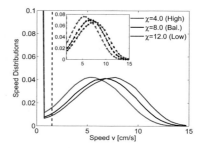

Figure 8.7.: Speed probability distributions of individuals for different diets (red: high protein; black: balanced; blue: low protein) obtained from experiments for $t > 300min$ and (b) from simulations of the constant rate model for different response strengths (red: $\chi = 4.0$; black: $\chi_{bal} = 8.0$; blue: $\chi_{low} = 12.0$). The dashed line represents $s = 1.5\ cm/s$. The insets show the corresponding speed probability distributions of only moving individuals (which we define as $s > 1.5\ cm/s$).

If our estimates of the social interaction strength, χ, are indeed associated with the respective diets, as our results suggest, it is possible to use the model to estimate the impact of the different diets on the onset of collective migration by varying the density for the corresponding interaction strengths.

The density was varied for different interaction strengths by keeping the number of individuals constant ($N = 576$) and changing the size of the simulation domain L. The onset of collective motion was measured using the global migration speed $\langle U \rangle$ and the corresponding order parameter Φ, where $\Phi = 0$ represents fully disordered motion and $\Phi = 1$ represents perfect alignment (collective motion):

$$\langle U \rangle = \left\langle \frac{1}{N} \left| \sum_{i=1}^{N} \mathbf{v}_i \right| \right\rangle \quad \text{and} \quad \Phi = \left\langle \frac{\left| \sum_{i=1}^{N} \mathbf{v}_i \right|}{\sum_{i=1}^{N} |\mathbf{v}_i|} \right\rangle, \tag{8.9}$$

with $\mathbf{v}_i$ being the velocity vector of individual i and $\langle \ldots \rangle$ denoting the average over 10 independent runs, each running for $2000s$ after the system reached a stationary state.

The simulations show that the mean speed and order parameter as a function of the density are strongly dependent on the interaction strengths (Fig. 8.8). An increase in density results in an increase in both the mean speed and order parameter for all interaction strengths. We define the critical density ρ_c for the onset of collective motion where the order parameter passes the value $\Phi = 0.5$[2]. For the constant transition rate model, the critical density decreases with increasing interaction strength, from $\rho_c^{(high)} \approx 80\text{m}^{-2}$ for χ_{high}, to $\rho_c^{(bal)} \approx 58\text{m}^{-2}$ for χ_{bal} and finally to $\rho_c^{(low)} \approx 52\text{m}^{-2}$ for χ_{low} (Fig. 8.8a,b). In

---

[2]The critical density for the onset of the ordered state in similar model systems is usually defined as the density at which the order parameter Φ starts to increase from $\Phi = 0$. In a finite system this point is difficult to estimate due to the smoothness of $\Phi(\rho)$ and we chose a more convenient definition

social force dependent transition rates model the following critical densities were obtained: $\rho_c^{(high)} \approx 65\mathrm{m}^{-2}$, $\rho_c^{(bal)} \approx 44\mathrm{m}^{-2}$ and $\rho_c^{(low)} \approx 38\mathrm{m}^{-2}$ (Fig. 8.8c,d). Whereas for the model variant with diet dependent rates the values are: $\rho_c^{(high)} \approx 80\mathrm{m}^{-2}$, $\rho_c^{(bal)} \approx 54\mathrm{m}^{-2}$ and $\rho_c^{(low)} \approx 45\mathrm{m}^{-2}$ (Fig. 8.8e,f).

In general, stronger escape and pursuit behaviour – associated with larger protein deprivation – leads to mass migration occurring at a much lower density. The qualitative differences in critical densities for different χ (diets) holds for the different model variants used and also different parameter values, but the quantitative results may differ. In particular the larger relative difference of the critical density between the high-protein and the balanced diet in comparison to the difference between the balanced and the low-protein diet appears to be an universal feature.

In summary, using this modelling ansatz based on experimental results, it is possible to establish a link between the nutritional state of individual locusts and the onset of collective motion in locusts swarms. Our results suggest that an abundant protein supply may supress the onset of collective motion at low and intermediate density of individuals. This prediction can, in principle, be tested in field studies on locusts in their natural environment.

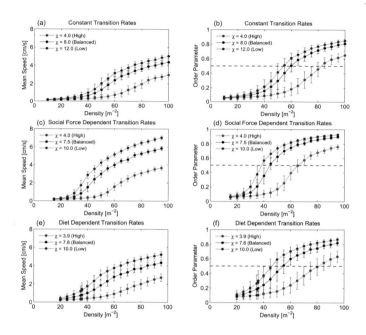

Figure 8.8.: Mean speed and order parameter for different transition rate models. The global migration speed (left panels) and the order parameter (right panels) as a function of density for individuals with different interaction strengths χ calculated from the model with a constant transition rate (a-b), transition rates dependent on the social interaction strength (c-d), and with diet dependent transition rates (e-f). Error bars show +/- one S. D. The dashed line on the right panels (b, d, f) indicates the threshold of the order parameter $\Phi = 0.5$ for the definition of the critical density. Each point was calculated as an average over time series from 10 independent simulations. The time series of each run was recorded over a time interval $\Delta t = 2000$ after the system reached a stationary state (numerical time step $dt = 0.006$).

9. Summary, Discussion and Outlook

In the first part of this work, I have focused on the dynamics of individual active particles. I discussed different models of active Brownian motion and derived analytical expressions for observables accessible in experiments such as stationary velocity (speed) probability distributions and the mean squared displacement. A major results that should be emphasized here is the distinction of active and passive fluctuations in the dynamics of active particles. I have shown that active velocity fluctuations may lead to divergence of the corresponding Cartesian velocity distributions and may have a strong impact on the effective coefficient of spatial diffusion of active particles with inertia.

Furthermore in the first part I have analyzed also some simple models of metabolism and addressed the question of optimal speed in animal motion. I hope the obtained results can be used in the future to establish a link between the dispersal and migration behavior of individuals, their internal metabolic dynamics and ecological factors such as nutrient distributions.

The second part of this work deals with collective dynamics of interacting active particles. I derived systematically a kinetic theory of active Brownian particles with a velocity alignment interaction. The obtained results are in general in a good agreement with the numerical simulations of the microscopic dynamics, but a detailed comparison revealed systematic deviations, such as for example the instability of the disordered mode in the one-dimensional Rayleigh-Helmholtz model, which can be understood from the approximations involved in the derivations of the mean field equations. The analysis was restricted to the simplest case of spatially homogeneous systems, but it is known from such active systems that they may lead to emergence of density inhomogeneities which may change the nature of the transition from disordered state (no collective motion) to the ordered state (collective motion). Thus in the future the question of the stability of the homogeneous state of collective motion should be addressed in this framework. The mean field theory was derived for external (Cartesian) noise terms, but the consideration of active fluctuations seems reasonable for future research. Preliminary considerations of active fluctuations indicate that the systematic derivation of the corresponding mean field equations is far from being trivial and therefore poses an interesting challenge for future research.

Furthermore, I introduced a model of individuals interacting via escape and pursuit responses. The model was motivated by recent experimental results which provide empirical evidence for cannibalism being the driving force of collective migration in certain insect species such as Mormon crickets (*Anabrus simplex*) and desert locusts (*Schistocerca gregaria*). It was shown that the escape and pursuit interaction leads to the onset of collective motion at high densities irrespective of detailed model parameters, whereas at low densities the behavior of the system is strongly dependent on the relative strength of individual escape and pursuit responses. Furthermore the scaling of the mean velocity of collective motion with the model parameters could be derived.

I presented also briefly a generalization of the escape and pursuit model for self-propelled particles with selective repulsion and attraction to other individuals in dependence on their relative velocity to the focal individual (approach versus escape). In particular, it was shown that collective motion in the generalized model appears only in the parameter space corresponding to the original escape-pursuit model. However, in addition to the collective motion it may account also for different individual behaviors such as attraction to everyone, which leads to disordered mosquito-like swarms, or repulsion from everyone which results in a maximal dispersal of individuals.

Finally, a modified escape and pursuit model was introduced, which accounts for the experimentally observed movement patterns of individual desert locusts. Based on the model simulations, we were able to show that the experimentally observed differences in the behavior of locusts in a group under different nutritional conditions, which are not observed for solitary individuals, can be explained via a nutritional state dependent social interaction. The simulations of the model parameterized by experimental data allow us to make predictions on the impact of the nutritional state on the critical density for the onset of collective migration in locusts.

The escape and pursuit framework introduced here is an interesting alternative to the models usually employed in the description of collective motion in biology based on attraction, velocity alignment and short range repulsion. It requires only local information which can easily be processed by individuals with restricted cognitive abilities, and shows a wide range of different patterns of collective motion. The possibility to "tune" the degree of spatial inhomogeneities in collective motion by changing the relative strength of escape and pursuit allows the analysis their impact on the phase transition between disordered and ordered states, which is an intensively debated subject in the field of statistical physics of active systems.

In the context of my work, I would like to close this final chapter with an appeal for a strong interdisciplinary collaboration between biologist and statistical physicist with an emphasis on a true two way exchange on equal terms. In the last decades we have already seen a strong increase in corresponding interdisciplinary research also beyond the classical example of biophysics. Based on my experience collaborating with biologists I am deeply convinced that the continuation and improvement of such links together with formation of new connections between the two disciplines will lead to further substantial advances in both.

Statistical physicists posses a large stock of methods for description and analysis of complex systems, which is being continuously developed, together with a long experience in applying this methods to natural phenomena. However, this in itself may be of limited value if applied in the biological context as there are simply too many questions which can be asked. In my opinion, a close collaboration with biologists is very helpful in order to address not only the "right" questions from the physics perspective, but also to be open to the important questions in the biological context. The question of "How?" often predominately addressed by physicist has to be complemented by the question of "Why?" from the biological perspective in order to achieve major breakthroughs.

On the other hand, for any answer to a "Why?"–question in biology, there may exist many different answers to the corresponding "How?"–question. The experience and understanding of complex systems by a statistical physicist can be of great advantage for finding the most satisfactory one.

This general idea may strike the reader as obvious and there are of course many scientists across the world from both disciplines who are aware of the great advantage of such scientific cross-fertilization. The fascinating research originating from their interdisciplinary efforts gives the best proof for that. Nevertheless it appears that despite common research interest the exchange of perspectives and ideas is often neglected and overcoming this tendency requires continuous effort from both sides.

Appendices

A. Speed model - Rescaling of the different propulsion function

In this appendix, we rescale the propulsion functions of active motion by rescaling the speed $\tilde{s}$ by the stationary speed s_0.

A.1. Schienbein-Gruler propulsion function

The Schienbein-Gruler speed model in the normal form reads:

$$\dot{\tilde{s}} = \alpha \left(1 - \frac{\tilde{s}}{s_0}\right) \tilde{s} + \sqrt{2\tilde{D}_s}, \xi \tag{A.1}$$

where $\tilde{s}$ is speed with dimension [length/time] and s_0 is the stationary speed. Introducing the dimensionless speed $s = \tilde{s}/s_0$ gives us $\tilde{s} = ss_0$ and transforms the above equation of motion to

$$\dot{s} = \alpha \left(1 - s\right) + \sqrt{2D_s}\xi, \tag{A.2}$$

with $D_s = \tilde{D}_s/v_0^2$ being the rescaled noise intensity.

A.2. Rayleigh-Helmholtz propulsion function

The Rayleigh-Helmholtz speed model in normal form reads

$$\dot{\tilde{s}} = (\gamma_1 - \gamma_2\tilde{s}^2)s + \sqrt{2\tilde{D}_s}\xi. \tag{A.3}$$

The stationary speed is determined by the coefficients of the RH-propulsion function $s_0 = \sqrt{\gamma_1/\gamma_2}$. Introducing the rescaled speed $s = \tilde{s}/s_0$ gives us

$$\dot{\tilde{s}} = \gamma_1(s - s^3) + \sqrt{2D_s}\xi, \tag{A.4}$$

with $D_s = \tilde{D}_s/s_0^2 = \tilde{D}_s\gamma_2/\gamma_1$ being the rescaled noise intensity. Throughout the chapter 4 we use $\gamma_1 \equiv \alpha$.

A.3. Schweitzer-Ebeling-Tilch propulsion function

The nonlinear propulsion function obtained from a limiting case of the depot model (Schweitzer et al., 1998) reads

$$\dot{\tilde{s}} = -\gamma_0 \tilde{s} + \frac{d_2 q_0 \tilde{s}}{c + d_2 \tilde{s}^2} + \sqrt{2\tilde{D}_s}\xi, \tag{A.5}$$

with γ_0 being the Stokes friction coefficient and d_2 (dissipation coefficient), q_0 (uptake coefficient), c (conversion coefficient) being the parameters of the energy depot. The stationary speed reads

$$s_0^2 = \frac{q_0}{\gamma_0} - \frac{c}{d_2}. \tag{A.6}$$

Rescaling the speed $\tilde{s}$ by the stationary speed s_0, corresponds to the substitution $\tilde{s} = s s_0$, with s being the scaled speed. It yields:

$$\dot{s} = \gamma_0 \left(\frac{\frac{q_0}{\gamma_0}}{\frac{c}{d_2} + \left(\frac{q_0}{\gamma_0} - \frac{c}{d_2} \right) s^2} - 1 \right) s + \sqrt{2D_s}\xi \tag{A.7}$$

$$= \gamma_0 \left(\frac{\frac{q_0 d_2}{c \gamma_0}}{1 + \left(\frac{q_0 d_2}{c \gamma_0} - 1 \right) s^2} - 1 \right) s + \sqrt{2D_s}\xi \tag{A.8}$$

$$= \gamma_0 \left(\frac{\beta}{1 + (\beta - 1) s^2} - 1 \right) s + \sqrt{2D_s}\xi, \tag{A.9}$$

with $\beta = q_0 d_2/(c\gamma_0)$ and $D_s = \tilde{D}_s/s_0^2$ being again the rescaled noise intensity. We use $\gamma_0 \equiv \alpha$ throughout the Chapter 4 .

B. Speed model - Stationary distributions and Moments

In this appendix we give the analytical expressions for the stationary speed distributions, as well as for the first two moments $\langle s \rangle$, $\langle s^2 \rangle$. The corresponding integrals were solved using the mathematical software MATHEMATICA 7.0 (Wolfram Research, 2008).

B.1. Schienbein-Gruler propulsion function

The stationary speed distribution for the Schienbein-Gruler speed model

$$\dot{s} = \alpha(1 - s) + \sqrt{2D_s}\xi, \tag{B.1}$$

with reflecting boundary at $s = 0$ reads

$$p_{SG}(s) = \mathcal{N}_{SG} \, \exp\left(-\frac{\alpha(s-1)^2}{2D_s}\right). \tag{B.2}$$

The normalization factor is given as

$$\mathcal{N}_{SG}^{-1} = \sqrt{\frac{\pi D_s}{2\alpha}} \exp\left(\frac{\alpha}{2D_s}\right) \left[1 + \mathrm{Erf}\left(\sqrt{\frac{\alpha}{2D_s}}\right)\right]. \tag{B.3}$$

The first two speed moments are

$$\langle s \rangle = 1 + \sqrt{\frac{2D_s}{\pi\alpha}} \frac{\exp\left(-\frac{\alpha}{2D_s}\right)}{1 + \mathrm{Erf}\left(\sqrt{\frac{\alpha}{2D_s}}\right)}, \tag{B.4}$$

$$\langle s^2 \rangle = 1 + \frac{D_s}{\alpha} + \sqrt{\frac{2D_s}{\pi\alpha}} \frac{\exp\left(-\frac{\alpha}{2D_s}\right)}{1 + \mathrm{Erf}\left(\sqrt{\frac{\alpha}{2D_s}}\right)}. \tag{B.5}$$

B.2. Rayleigh-Helmholtz propulsion function

The stationary speed distribution for the Rayleigh-Helmholtz speed model

$$\dot{s} = \alpha(s - s^3) + \sqrt{2D_s}\xi, \tag{B.6}$$

with reflecting boundary at $s = 0$ reads

$$p_{RH}(s) = \mathcal{N}_{RH} \ \exp\left(-\frac{\alpha s^2}{2D_s} + \frac{\alpha s^4}{4D_s}\right). \tag{B.7}$$

The normalization factor reads

$$\mathcal{N}_{RH}^{-1} = \frac{\pi}{4} \exp\left(\frac{\alpha}{8D_s}\right) \left[I_{-\frac{1}{4}}\left(\frac{\alpha}{8D_s}\right) + I_{\frac{1}{4}}\left(\frac{\alpha}{8D_s}\right)\right], \tag{B.8}$$

with $I_n(x)$ being the modified Bessel function of the first kind.

The first two speed moments are

$$\langle s \rangle = \mathcal{N}_{RH} \ \sqrt{\frac{\pi D_s}{4\alpha}} \exp\left(\frac{\alpha}{4D_s}\right) \left[1 + \mathrm{Erf}\left(\sqrt{\frac{\alpha}{4D_s}}\right)\right], \tag{B.9}$$

$$\langle s^2 \rangle = \mathcal{N}_{RH} \left[\frac{\left(\frac{D_s}{\alpha}\right)^{\frac{3}{4}} \pi \ {}_1F_1\left(\frac{3}{4}, \frac{1}{2}, \frac{\alpha}{4D_s}\right)}{\Gamma(\frac{1}{4})} - \frac{\left(\frac{D_s}{\alpha}\right)^{\frac{1}{4}} \pi \ {}_1F_1\left(\frac{5}{4}, \frac{3}{2}, \frac{\alpha}{4D_s}\right)}{\Gamma(-\frac{1}{4})}\right], \tag{B.10}$$

where ${}_1F_1(a, b, c)$ represents Kummer's confluent hypergeometric function.

B.3. Approximated Schweitzer-Ebeling-Tilch propulsion function

The stationary speed distribution for the approximated Schweitzer-Ebeling-Tilch speed model (aSET)

$$\dot{s} = \alpha \left(\frac{1}{s} - s\right) + \sqrt{2D_s}\xi, \tag{B.11}$$

with reflecting boundary at $s = 0$ reads

$$p_{aSET}(s) = \mathcal{N}_{aSET} \ s^{\frac{\alpha}{D_s}} \exp\left(-\frac{\alpha s^2}{2D_s}\right). \tag{B.12}$$

The normalization factor reads

$$\mathcal{N}_{aSET}^{-1} = \frac{1}{2} \left(\frac{\alpha}{2D_s}\right)^{-\frac{\alpha + D_s}{2D_s}} \Gamma\left(\frac{\alpha + D_s}{2D_s}\right). \tag{B.13}$$

The first two speed moments are

$$\langle s \rangle = \sqrt{\frac{\alpha}{2D_s}} \frac{\Gamma\left(\frac{\alpha}{2D_s}\right)}{\Gamma\left(\frac{\alpha + D_s}{2D_s}\right)}, \tag{B.14}$$

$$\langle s^2 \rangle = 1 + \frac{D_s}{\alpha}. \tag{B.15}$$

C. Numerical Methods and Software

The main programming language used for numerical simulations was (ANSI) C supplemented with additional routines from different libraries, in particular, the GNU Scientific Library (Galassi et al., 2009).

The numerical simulations of individual noninteracting particles were performed using the Heun algorithm, which naturally yields the Stratonovich interpretation of the stochachastic integral. For interacting particles (only additive noise), the simple Euler scheme has been used. A review and analysis of the different algorithms can be found in Mannella (2000).

The time step in numerical simulations was chosen according to the required accuracy in the range $\Delta t = 10^{-2} - 10^{-5}$.

A significant number of numerical simulations was performed on graphical processing units (GPUs) using CUDA (NVIDIA, 2008) (Computing Unified Device Architecture). The intrinsically parallel hardware structure of GPUs, allows massive speed-up of general purpose numerical calculations. The massive acceleration of numerical simulations of stochastic differential equations using CUDA has been discussed in detail by Januszewski and Kostur (2010). Their publication includes references to corresponding source code samples and is well suited for introductionary reading on general purpose computing using CUDA.

Complex analytical transformations and solutions of certain integrals were either checked or obtained using MATHEMATICA® 7.0 (Wolfram Research, 2008). The analysis and processing of simulation and experimental data was has been performed using MATLAB 7.10® (MathWorks, 2010).

Nomenclature

Δl_w .... distance to the wall of the experimental arena of individual locusts, page 113

δv_k .... deviations of k-th component of individual velocity from the mean field velocity, page 68

$\gamma(v)$ .... velocity dependent friction function, page 9

γ_0 .... friction coefficient (Stokes friction), page 90

κ_i .... transition rates of the dichotomous Markov process, page 11

$\kappa_{v,\varphi}$ .... relaxation rates of velocity and angular variable, page 33

$\lambda(s_0)$ .... speed dependent distance an active agent can cover before depletion of internal energy, page 53

$\lambda_{\max}$ .... maximal distance an active agent may travel before depletion of internal energy depot moving at optimal speed, page 54

μ .... velocity-alignment strength, page 66

μ_a .... response strength to approaching individuals, page 105

μ_m .... response strength to moving away individuals, page 105

μ_r .... short range repulsion strength, page 105

Φ .... order parameter of collective motion, page 106

ρ .... density (number of particles per unit area), page 66

ρ_s .... dimensionless density scaled by the interaction range of individuals, page 92

σ^2 .... variance of a probability distribution, page 10

τ_v .... velocity correlation time, page 9

θ .... squared temperature fluctuations, page 68

ε .... interaction range for the velocity alignment interaction, page 66

φ .... angular variable defining heading direction, page 19

$\xi(t)$ .... white Gaussian random variable, page 10

D .... (external) noise intensity, page 10

$\mathcal{C}$ .... (internal) energy conversion function, page 51

c .... conversion factor of internal energy into kinetic energy, page 52

l_r .... hardcore or short range repulsion diameter, page 90

l_s .... sensory range of individuals (interaction range, escape+pursuit), page 90

m .... particle mass, page 9

$M_k^{(n)}$ .... n-th moment of the of the velocity component k, page 66

$M_{kl}^{(nm)}$ .... mixed moments (nm) of the of the velocity component kl, page 67

$\mathcal{N}$ .... normalization constant (probability density distribution), page 10

N .... number of particles, page 66

N_ε .... number of particles within the interaction range of particle i, page 66

$N_{e,p,r}$ .... number of individuals the focal individual responds to with escape (e), pursuit (p) or short range repulsion (r), page 91

$p_s(x)$ .... stationary probability distribution of variable x, page 10

$\mathcal{Q}$ .... (internal) energy uptake function, page 51

q_0 .... (internal) energy uptake rate, page 52

$q_{+/-}$ .... state variables of the dichotomous Markov process, page 11

R_{hc} .... hardcore or short range repulsion radius, page 90

$\mathbf{r}$ .... position vector, page 19

$\mathbf{r}_{ji}$ .... distance vector pointing from individual i to individual j, page 91

$\langle S \rangle$ .... time averaged migration speed in "Move-or-Forage" model, page 56

s .... speed of an active particle, page 39

s_{opt} .... optimal speed of individual active agents (e.g. speed maximizing the distance travelled or the mean migration speed), page 53

t .... time, page 9

T_k .... mean field temperature component (squared velocity deviations), page 68

$T_{\perp,\parallel}$ .... temperature components parallel and perpedicular to the mean field direction of motion, page 76

Bibliography

N. Abaid and M. Porfiri. Fish in a ring: spatio-temporal pattern formation in one-dimensional animal groups. *Journal of The Royal Society Interface*, 2010. ISSN 1742-5689.

M. Aldana, V. Dossetti, C. Huepe, V. M. Kenkre, and H. Larralde. Phase transitions in systems of Self-Propelled agents and related network models. *Physical Review Letters*, 98(9):095702–4, 2007.

T. Alerstam, A. Hedenström, and S. Åkesson. Long-distance migration: evolution and determinants. *Oikos*, 103(2):247–260, 2003. ISSN 1600-0706.

I. Aoki. A simulation study on the schooling mechanism in fish. *Bulletin of the Japanese Society of Scientific Fisheries (Japan)*, 1982.

M. Ballerini, N. Cabibbo, R. Candelier, A. Cavagna, E. Cisbani, I. Giardina, V. Lecomte, A. Orlandi, G. Parisi, A. Procaccini, M. Viale, and V. Zdravkovic. Interaction ruling animal collective behavior depends on topological rather than metric distance: Evidence from a field study. *Proceedings of the National Academy of Sciences*, 105(4):1232 –1237, Jan. 2008. doi: 10.1073/pnas.0711437105.

F. Bartumeus, M. G. E. da Luz, G. M. Viswanathan, and J. Catalan. Animal search strategies: A quantitative Random-Walk analysis. *Ecology*, 86(11):3078–3087, 2005. ISSN 0012-9658. doi: 10.1890/04-1806.

F. Bartumeus, J. Catalan, G. Viswanathan, E. Raposo, and M. da Luz. The influence of turning angles on the success of non-oriented animal searches. *Journal of Theoretical Biology*, 252(1):43–55, 2008. ISSN 0022-5193. doi: 10.1016/j.jtbi.2008.01.009.

A. Baskaran and M. C. Marchetti. Statistical mechanics and hydrodynamics of bacterial suspensions. *Proceedings of National Academy of Science USA*, 106(37):15567–15572, Sept. 2009. ISSN 0027-8424. doi: 10.1073/pnas.0906586106. PMID: 19717428 PMCID: 2747162.

S. Bazazi, J. Buhl, J. J. Hale, M. L. Anstey, G. A. Sword, S. J. Simpson, and I. D. Couzin. Collective motion and cannibalism in locust migratory bands. *Current Biology*, 18(10): 735–739, 2008. ISSN 0960-9822. doi: 10.1016/j.cub.2008.04.035.

S. Bazazi, P. Romanczuk, S. Thomas, L. Schimansky-Geier, J. J. Hale, G. A. Miller, G. A. Sword, S. J. Simpson, and I. D. Couzin. Nutritional state and collective motion: from individuals to mass migration. *Proceedings of the Royal Society B: Biological Sciences*, 278(1704):356–363, 2011. ISSN 0962-8452. doi: 10.1098/rspb.2010.1447.

E. Bertin, M. Droz, and G. Gregoire. Boltzmann and hydrodynamic description for self-propelled particles. *Physical Review E (Statistical, Nonlinear, and Soft Matter Physics)*, 74(2):022101–4, 2006.

H. U. Bödeker, C. Beta, T. D. Frank, and E. Bodenschatz. Quantitative analysis of random ameboid motion. *EPL (Europhysics Letters)*, 90(2):28005, 2010. ISSN 0295-5075. doi: 10.1209/0295-5075/90/28005.

N. V. Brilliantov and T. Pöschel. *Kinetic theory of granular gases*. Oxford Univ.Press, 2004.

R. Brown. XXVII. a brief account of microscopical observations made in the months of june, july and august 1827, on the particles contained in the pollen of plants; and on the general existence of active molecules in organic and inorganic bodies. *Philosophical Magazine Series 2*, 4(21):161–173, 1828. ISSN 1941-5850.

J. Buhl, D. J. T. Sumpter, I. D. Couzin, J. J. Hale, E. Despland, E. R. Miller, and S. J. Simpson. From disorder to order in marching locusts. *Science*, 312(5778):1402–1406, June 2006. doi: 10.1126/science.1125142.

H. Chaté, F. Ginelli, and R. Montagne. Simple model for active nematics: Quasi-Long-Range order and giant fluctuations. *Physical Review Letters*, 96(18):180602–4, 2006.

H. Chaté, F. Ginelli, G. Grégoire, and F. Raynaud. Collective motion of self-propelled particles interacting without cohesion. *Physical Review E*, 77(4):046113, Apr. 2008. doi: 10.1103/PhysRevE.77.046113.

I. D. Couzin, J. Krause, R. James, G. D. Ruxton, and N. R. Franks. Collective memory and spatial sorting in animal groups. *Journal of Theoretical Biology*, 218(1):1–11, Sept. 2002. doi: 10.1006/jtbi.2002.3065.

A. Czirók, E. Ben-Jacob, I. Cohen, and T. Vicsek. Formation of complex bacterial colonies via self-generated vortices. *Physical Review E*, 54(2):1791, 1996. doi: 10.1103/PhysRevE.54.1791.

J. Deseigne, O. Dauchot, and H. Chaté. Collective motion of vibrated polar disks. *Physical Review Letters*, 105(9):098001, 2010. doi: 10.1103/PhysRevLett.105.098001.

E. J. Doedel. AUTO: a program for the automatic bifurcation analysis of autonomous systems. *Congr. Numer*, 30:265–284, 1981.

M. R. D'Orsogna, Y. L. Chuang, A. L. Bertozzi, and L. S. Chayes. Self-Propelled particles with Soft-Core interactions: Patterns, stability, and collapse. *Physical Review Letters*, 96:104302, 2006.

B. Dybiec and L. Schimansky-Geier. Emergence of bimodality in noisy systems with single-well potential. *The European Physical Journal B - Condensed Matter and Complex Systems*, 57(3):313–320, June 2007. doi: 10.1140/epjb/e2007-00162-y.

W. Ebeling and L. Schimansky-Geier. Swarm dynamics, attractors and bifurcations of active Brownian motion. *The European Physical Journal - Special Topics*, 157(1):17–31, Apr. 2008. doi: 10.1140/epjst/e2008-00627-9.

W. Ebeling and I. M. Sokolov. *Statistical thermodynamics and stochastic theory of nonequilibrium systems*. World Scientific, 2005. ISBN 9789810213824.

W. Ebeling, F. Schweitzer, and B. Tilch. Active Brownian particles with energy depots modelling animal mobility. *BioSystems*, 49:17–29, 1999.

A. Einstein. The motion of elements suspended in static liquids as claimed in the molecular kinetic theory of heat. *Annalen der Physik*, 17(8):549–560, July 1905. ISSN 0003-3804.

U. Erdmann. *Kollektive Bewegung: Komplexe Strukturen in 2D-Systemen aktiver Brown'scher Teilchen fernab vom Gleichgewicht*, volume 10 of *Nichtlineare und Stochastische Physik*. Logos-Verlag, Berlin, 2003.

U. Erdmann and W. Ebeling. Collective motion of Brownian particles with hydrodynamic interactions. *Fluctuation and Noise Letters*, 3(2):L145–L154, 2003.

U. Erdmann and S. Göller. Aktive Teilchen auf der Futtersuche. In *Irreversible Prozesse und Selbstorganisation*. Logos Verlag, Berlin, 2006.

U. Erdmann, W. Ebeling, L. Schimansky-Geier, and F. Schweitzer. Brownian particles far from equilibrium. *The European Physical Journal B - Condensed Matter and Complex Systems*, 15(1):105–113, Apr. 2000. doi: 10.1007/s100510051104.

U. Erdmann, W. Ebeling, and V. S. Anishchenko. Excitation of rotational modes in two-dimensional systems of driven Brownian particles. *Phys. Rev. E*, 65(6):061106, June 2002.

U. Erdmann, W. Ebeling, and A. S. Mikhailov. Noise induced transition from translational to rotational motion of swarms. *Physical Review E*, 71(5):051904, 2005.

B. Ermentrout. *Simulating, Analyzing, and Animating Dynamical Systems: A Guide to Xppaut for Researchers and Students (Software, Environments, Tools)*. Society for Industrial Mathematics, 1st edition, Mar. 2002. ISBN 0898715067.

M. A. Fedak, N. C. Heglund, and C. R. Taylor. Energetics and mechanics of terrestrial locomotion. II. kinetic energy changes of the limbs and body as a function of speed and body size in birds and mammals. *The Journal of Experimental Biology*, 97:23–40, Apr. 1982. ISSN 0022-0949. PMID: 7086342.

B. M. Friedrich and F. Julicher. Chemotaxis of sperm cells. *Proceedings of the National Academy of Sciences*, 104(33):13256, 2007.

R. Fürth. Die Brownsche Bewegung bei Berücksichtigung einer Persistenz der Bewegungsrichtung. Mit Anwendungen auf die Bewegung lebender Infusorien. *Zeitschrift für Physik*, 2(3):244–256, 1920. ISSN 1434-6001. doi: 10.1007/BF01328731.

M. Galassi, J. Davies, J. Theiler, B. Gough, G. Jungman, P. Alken, M. Booth, and F. Rossi. *GNU Scientific Library Reference Manual*. Network Theory Ltd, 3 edition, 2009. ISBN 0-9546120-7-8.

C. W. Gardiner. *Handbook of Stochastic Methods for Physics, Chemistry and the Natural Sciences*, volume 13 of *Springer Series in Synergetics*. Springer, Berlin Heidelberg New York, 2. edition, 1985.

G. Grégoire and H. Chaté. Onset of collective and cohesive motion. *Physical Review Letters*, 92(2):025702, 2004.

D. Grossman, I. S. Aranson, and E. B. Jacob. Emergence of agent swarm migration and vortex formation through inelastic collisions. *New Journal of Physics*, 10(023036): 023036, 2008.

V. Guttal and I. D. Couzin. Social interactions, information use, and the evolution of collective migration. *Proceedings of the National Academy of Sciences*, 107(37):16172–16177, 2010. doi: 10.1073/pnas.1006874107.

K. P. Hadeler. Reaction telegraph equations and random walk systems. *Stochastic and spatial structures of dynamical systems*, page 133, 1996.

N. Heglund, M. Fedak, C. Taylor, and G. Cavagna. Energetics and mechanics of terrestrial locomotion. IV. total mechanical energy changes as a function of speed and body size in birds and mammals. *J Exp Biol*, 97(1):57–66, Apr. 1982a.

N. C. Heglund, G. A. Cavagna, and C. R. Taylor. Energetics and mechanics of terrestrial locomotion. III. energy changes of the centre of mass as a function of speed and body size in birds and mammals. *The Journal of Experimental Biology*, 97:41–56, Apr. 1982b. ISSN 0022-0949. PMID: 7086349.

J. P. Hernandez-Ortiz, C. G. Stoltz, and M. D. Graham. Transport and collective dynamics in suspensions of confined swimming particles. *Physical Review Letters*, 95(20):204501–4, Nov. 2005.

T. Hillen and H. G. Othmer. The diffusion limit of transport equations derived from Velocity-Jump processes. *SIAM Journal on Applied Mathematics*, 61(3):751–775, 2000. ISSN 00361399.

J. R. Howse, R. A. L. Jones, A. J. Ryan, T. Gough, R. Vafabakhsh, and R. Golestanian. Self-Motile colloidal particles: From directed propulsion to random walk. *Physical Review Letters*, 99(4):048102, July 2007. doi: 10.1103/PhysRevLett.99.048102.

D. F. Hoyt and C. R. Taylor. Gait and the energetics of locomotion in horses. *Nature*, 292(5820):239–240, July 1981. doi: 10.1038/292239a0.

A. Huth and C. Wissel. The simulation of the movement of fish schools. *Journal of Theoretical Biology*, 156(3):365–385, June 1992. ISSN 0022-5193. doi: 10.1016/S0022-5193(05) 80681-2.

J. Ingen-Housz. Bemerkungen über den Gebrauch des Vergrösserungsglases. *Verm. Schriften physisch-medicinischen Inhalts*, 1784.

M. Januszewski and M. Kostur. Accelerating numerical solution of stochastic differential equations with CUDA. *Computer Physics Communications*, 181(1):183–188, Jan. 2010. ISSN 0010-4655. doi: 10.1016/j.cpc.2009.09.009.

M. Kac. A stochastic model related to the telegrapher's equation.(1956), reprinted in rocky mtn. *Math. J*, 4:497–509, 1974.

P. M. Kareiva and N. Shigesada. Analyzing insect movement as a correlated random walk. *Oecologia*, 56(2-3):234–238, 1983. ISSN 0029-8549. doi: 10.1007/BF00379695.

N. Komin, U. Erdmann, and L. Schimansky-Geier. Random walk theory applied to daphnia motion. *Fluctuation and Noise Letters*, 4(1):L151–L159, 2004.

J. M. Kosterlitz. The critical properties of the two-dimensional xy model. *J. Phys. C*, 7: 1046, 1974.

J. M. Kosterlitz and D. J. Thouless. Ordering, metastability and phase transitions in two-dimensional systems. *J. Phys. C*, 6(1181):10, 1973.

R. Kram and C. R. Taylor. Energetics of running: a new perspective. *Nature,* 346(6281): 265–267, 1990. ISSN 0028-0836. doi: 10.1038/346265a0.

D. J. Krause and G. D. Ruxton. *Living in groups*. Oxford University Press, 2002. ISBN 9780198508182.

A. Kudrolli, G. Lumay, D. Volfson, and L. S. Tsimring. Swarming and swirling in Self-Propelled polar granular rods. *Physical Review Letters*, 100(5):058001, Feb. 2008. doi: 10.1103/PhysRevLett.100.058001.

A. D. Kuo. Energetics of actively powered locomotion using the simplest walking model. *Journal of Biomechanical Engineering*, 124(1):113–120, Feb. 2002. doi: 10.1115/1.1427703.

P. Langevin. The theory of Brownian movement. *Comptes Rendus Hebdomadaires des Seances de L'Academie des Sciences*, 146:530–533, 1908. ISSN 0001-4036.

A. W. Liehr, H. U. Bödeker, M. C. Röttger, T. D. Frank, R. Friedrich, and H. Purwins. Drift bifurcation detection for dissipative solitons. *New Journal of Physics*, 5:89–89, 2003. ISSN 1367-2630. doi: 10.1088/1367-2630/5/1/389.

B. Lindner. The diffusion coefficient of nonlinear Brownian motion. *New Journal of Physics*, 9(5):136–136, 2007. ISSN 1367-2630. doi: 10.1088/1367-2630/9/5/136.

B. Lindner and E. M. Nicola. Critical asymmetry for giant diffusion of active Brownian particles. *Physical Review Letters*, 101(19):190603, Nov. 2008a. doi: 10.1103/PhysRevLett.101.190603.

B. Lindner and E. M. Nicola. Diffusion in different models of active Brownian motion. *The European Physical Journal - Special Topics*, 157(1):43–52, Apr. 2008b. doi: 10.1140/epjst/e2008-00629-7.

R. Lukeman, Y. Li, and L. Edelstein-Keshet. Inferring individual rules from collective behavior. *Proceedings of the National Academy of Sciences*, 107(28):12576–12580, July 2010. doi: 10.1073/pnas.1001763107.

R. Mannella. A gentle introduction to the integration of stochastic differential equations. *Lecture Notes in Physics (Springer)*, 557:353, 2000.

MathWorks. *MATLAB version 7.10.0 (R2010a)*. The MathWorks Inc., Natnick, Massachusets, 2010.

A. S. Mikhailov and D. Meinköhn. Self-motion in physico-chemical systems far from equilibrium. In L. Schimansky-Geier and T. Pöschel, editors, *Stochastic Dynamics*, volume 484 of *Lecture Notes in Physics*, pages 334–345. Springer Berlin, 1997.

J. G. Mitchell. The energetics and scaling of search strategies in bacteria. *American Naturalist*, 160(6):727–740, 2002. ISSN 0003-0147.

M. Nagy, I. Daruka, and T. Vicsek. New aspects of the continuous phase transition in the scalar noise model (SNM) of collective motion. *Physica A: Statistical and Theoretical Physics*, 373:445–454, Jan. 2007.

H. Niwa. Self-organizing dynamic model of fish schooling. *Journal of Theoretical Biology*, 171(2):123–136, Nov. 1994. ISSN 0022-5193. doi: 10.1006/jtbi.1994.1218.

NVIDIA. *CUDA Programming Guide 2.0*. NVIDIA Corporation, Santa Clara, CA, USA, 2008.

A. Okubo and S. A. Levin. *Diffusion and Ecological Problems: Modern Perspectives*, volume 14 of *Interdisciplinary Applied Mathematics*. Springer, New York, 2. edition, 2001.

L. S. Ornstein. On the Brownian motion. *Koninklijke Nederlandse Akademie van Weteschappen Proceedings Series B Physical Sciences*, 21:96–108, 1919.

Othmer, Dunbar, and Alt. Models of dispersal in biological systems. *Journal of Mathematical Biology*, 26(3):263–298, June 1988. doi: 10.1007/BF00277392.

R. F. Pawula. Approximation of the linear boltzmann equation by the Fokker-Planck equation. *Physical Review*, 162(1):186, 1967. doi: 10.1103/PhysRev.162.186.

R. F. Pawula. Approximating distributions from moments. *Physical Review A*, 36(10): 4996, Nov. 1987. doi: 10.1103/PhysRevA.36.4996.

W. F. Paxton, K. C. Kistler, C. C. Olmeda, A. Sen, S. K. S. Angelo, Y. Cao, T. E. Mallouk, P. E. Lammert, and V. H. Crespi. Catalytic nanomotors: Autonomous movement of striped nanorods. *Journal of the American Chemical Society*, 126(41):13424–13431, 2004. doi: 10.1021/ja047697z.

F. Peruani and L. G. Morelli. Self-Propelled particles with fluctuating speed and direction of motion in two dimensions. *Physical Review Letters*, 99(1):010602, July 2007. doi: 10.1103/PhysRevLett.99.010602.

F. Peruani, A. Deutsch, and M. Bär. Nonequilibrium clustering of self-propelled rods. *Physical Review E (Statistical, Nonlinear, and Soft Matter Physics)*, 74(3):030904–4, 2006.

F. Peruani, A. Deutsch, and M. Bär. A mean-field theory for self-propelled particles interacting by velocity alignment mechanisms. *The European Physical Journal Special Topics*, 157(1):111–122, 2008. ISSN 1951-6355. doi: 10.1140/epjst/e2008-00634-x.

K. Przibram. Über die ungeordnete Bewegung niederer Tiere. *Pflüger, Archiv für die Gesammte Physiologie des Menschen und der Thiere*, 153(8-10):401–405, 1913. ISSN 0031-6768. doi: 10.1007/BF01686480.

K. Przibram. Über die ungeordnete bewegung niederer tiere. II. *Archiv für Entwicklungsmechanik der Organismen*, 43(1-2):20–27, 1917. ISSN 0949-944X. doi: 10.1007/BF02189255.

H. Ralston. Energy-speed relation and optimal speed during level walking. *European Journal of Applied Physiology and Occupational Physiology*, 17(4):277–283, Oct. 1958. doi: 10.1007/BF00698754.

S. Ramaswamy. The mechanics and statistics of active matter. *Annual Review of Condensed Matter Physics*, 1(1):323–345, Aug. 2010. ISSN 1947-5454. doi: 10.1146/annurev-conmatphys-070909-104101.

J. W. S. Rayleigh. *The Theory of Sound*, volume I. MacMillan, London, 2. edition, 1894. erstmalige Erwaehnung des sog. Rayleigh-Oszillators.

P. Reimann. Brownian motors: noisy transport far from equilibrium. *Physics Reports*, 361(2-4):57–265, Apr. 2002. ISSN 0370-1573. doi: 10.1016/S0370-1573(01)00081-3.

C. W. Reynolds. Flocks, herds and schools: A distributed behavioral model. In *Proceedings of the 14th annual conference on Computer graphics and interactive techniques - SIGGRAPH '87*, pages 25–34, 1987. doi: 10.1145/37401.37406.

T. Riethmüller, L. Schimansky-Geier, D. Rosenkranz, and T. Pöschel. Langevin equation approach to granular flow in a narrow pipe. *Journal of Statistical Physics*, 86(1):421–430, Jan. 1997. doi: 10.1007/BF02180213.

F. C. Rind and P. J. Simmons. Orthopteran DCMD neuron: a reevaluation of responses to moving objects. i. selective responses to approaching objects. *J Neurophysiol*, 68(5): 1654–1666, Nov. 1992.

F. C. Rind, R. D. Santer, and G. A. Wright. Arousal facilitates collision avoidance mediated by a looming sensitive visual neuron in a flying locust. *J Neurophysiol*, 100(2): 670–680, Aug. 2008. doi: 10.1152/jn.01055.2007.

H. Risken. *The Fokker-Planck equation: Methods of solution and applications*. Springer Verlag, 1996. ISBN 354061530X.

S. M. Rogers, G. W. J. Harston, F. Kilburn-Toppin, T. Matheson, M. Burrows, F. Gabbiani, and H. G. Krapp. Spatiotemporal receptive field properties of a Looming-Sensitive neuron in solitarious and gregarious phases of the desert locust. *J Neurophysiol*, 103(2): 779–792, Feb. 2010. doi: 10.1152/jn.00855.2009.

P. Romanczuk, U. Erdmann, H. Engel, and L. Schimansky-Geier. Beyond the Keller-Segel model. *The European Physical Journal Special Topics*, 157(1):61–77, 2008. ISSN 1951-6355. doi: 10.1140/epjst/e2008-00631-1.

P. Romanczuk, I. D. Couzin, and L. Schimansky-Geier. Collective motion due to individual escape and pursuit response. *Physical Review Letters*, 102(1):010602–4, Jan. 2009. doi: 10.1103/PhysRevLett.102.010602.

H. J. Rothe, W. Biesel, and W. Nachtigall. Pigeon flight in a wind tunnel. *Journal of Comparative Physiology B: Biochemical, Systemic, and Environmental Physiology*, 157 (1):99–109, Jan. 1987. doi: 10.1007/BF00702734.

G. Ruckner and R. Kapral. Chemically powered nanodimers. *Physical Review Letters*, 98 (15):150603–4, Apr. 2007.

I. Rushkin, V. Kantsler, and R. E. Goldstein. Fluid velocity fluctuations in a suspension of swimming protists. *Physical Review Letters*, 105(18):188101, 2010. doi: 10.1103/PhysRevLett.105.188101.

D. Saintillan and M. J. Shelley. Orientational order and instabilities in suspensions of Self-Locomoting rods. *Physical Review Letters*, 99(5):058102, July 2007. doi: 10.1103/PhysRevLett.99.058102.

F. D. D. Santos and T. Ondarçuhu. Free-Running droplets. *Physical Review Letters*, 75 (16):2972, 1995. doi: 10.1103/PhysRevLett.75.2972.

M. Schienbein and H. Gruler. Langevin equation, Fokker-Planck equation and cell migration. *Bulletin of Mathematical Biology*, 55(3):585–608, 1993. ISSN 0092-8240. doi: 10.1007/BF02460652.

L. Schimansky-Geier, M. Mieth, H. Rosè, and H. Malchow. Structure formation by active Brownian particles. *Physics Letters A*, 207:140–146, 1995.

L. Schimansky-Geier, U. Erdmann, and N. Komin. Advantages of hopping on a zig-zag course. *Physica A: Statistical Mechanics and its Applications*, 351(1):51–59, June 2005. ISSN 0378-4371. doi: 10.1016/j.physa.2004.12.043.

K. Schmidt-Nielsen. Locomotion: energy cost of swimming, flying, and running. *Science*, 177(45):222–228, 1972.

F. Schweitzer. *Brownian Agents and Active Particles: Collective Dynamics in the Natural and Social Sciences*. Springer, 1 edition, Aug. 2003. ISBN 3540439382.

F. Schweitzer, W. Ebeling, and B. Tilch. Complex motion of Brownian particles with energy depots. *Physical Review Letters*, 80(23):5044, June 1998. doi: 10.1103/PhysRevLett.80.5044.

D. Selmeczi. Cell motility as persistent random motion: Theories from experiments. *Biophysical Journal*, 89(2):912–931, 2005. ISSN 00063495. doi: 10.1529/biophysj.105.061150.

D. Selmeczi, L. Li, L. I. Pedersen, S. F. Nørrelykke, P. H. Hagedorn, S. Mosler, N. B. Larsen, E. C. Cox, and H. Flyvbjerg. Cell motility as random motion: A review. *The European Physical Journal Special Topics*, 157(1):1–15, 2008. ISSN 1951-6355. doi: 10.1140/epjst/e2008-00626-x.

R. A. Simha and S. Ramaswamy. Hydrodynamic fluctuations and instabilities in ordered suspensions of Self-Propelled particles. *Physical Review Letters*, 89(5):058101, 2002a.

R. A. Simha and S. Ramaswamy. Statistical hydrodynamics of ordered suspensions of self-propelled particles: waves, giant number fluctuations and instabilities. *Physical A*, 306:262–269, 2002b.

S. J. Simpson, G. A. Sword, P. D. Lorch, and I. D. Couzin. Cannibal crickets on a forced march for protein and salt. *PNAS*, 103(11):4152–4156, Mar. 2006.

A. R. Sinclair and P. Arcese, editors. *Serengeti II: dynamics, management, and conservation of an ecosystem.* University of Chicago Press, 1995. ISBN 0226760324.

M. D. Sockol, D. A. Raichlen, and H. Pontzer. Chimpanzee locomotor energetics and the origin of human bipedalism. *Proceedings of the National Academy of Sciences*, 104(30): 12265–12269, July 2007. doi: 10.1073/pnas.0703267104.

A. Sokolov, I. S. Aranson, J. O. Kessler, and R. E. Goldstein. Concentration dependence of the collective dynamics of swimming bacteria. *Physical Review Letters*, 98(15):158102–4, Apr. 2007.

A. Sokolov, R. E. Goldstein, F. I. Feldchtein, and I. S. Aranson. Enhanced mixing and spatial instability in concentrated bacterial suspensions. *Physical Review E*, 80(3):031903, 2009. doi: 10.1103/PhysRevE.80.031903.

J. Strefler. *Swarming Theory in Three Dimensions based on Active Brownian Particles.* Diploma thesis, Humboldt Universität zu Berlin, 2007.

J. Strefler, U. Erdmann, and L. Schimansky-Geier. Swarming in three dimensions. *Physical Review E*, 78(3):031927, 2008. doi: 10.1103/PhysRevE.78.031927.

Y. Sumino, N. Magome, T. Hamada, and K. Yoshikawa. Self-Running droplet: Emergence of regular motion from nonequilibrium noise. *Physical Review Letters*, 94(6):068301–4, Feb. 2005.

R. Suzuki and S. Sakai. Movement of a group of animals. *Biophysics*, 13:281–282, 1973.

C. R. Taylor and N. C. Heglund. Energetics and mechanics of terrestrial locomotion. *Annual review of physiology*, 44(1):97–107, 1982. ISSN 0066-4278.

P. Tierno, R. Albalat, and F. Sagués. Autonomously moving catalytic microellipsoids dynamically guided by external magnetic fields. *Small*, 6(16):1749–1752, 2010. ISSN 16136810. doi: 10.1002/smll.201000832.

J. Toner and Y. Tu. Long-Range order in a Two-Dimensional dynamical XY model: How birds fly together. *Physical Review Letters*, 75(23):4326, 1995. doi: 10.1103/PhysRevLett.75.4326.

J. Toner and Y. Tu. Flocks, herds, and schools: A quantitative theory of flocking. *Physical Review E*, 58(4):4828–4858, 1998.

V. A. Tucker. Energetic cost of locomotion in animals. *Comparative Biochemistry and Physiology*, 34(4):841–846, June 1970. ISSN 0010-406X. doi: 10.1016/0010-406X(70) 91006-6.

N. G. van Kampen. *Stochastic Processes in Physics and Chemistry*. North-Holland, Amsterdam, 2. edition, 1992.

T. Vicsek, A. Czirók, E. Ben-Jacob, I. Cohen, and O. Shochet. Novel type of phase transisition in a system of Self-Driven particles. *Physical Review Letters*, 75(6):1226–1229, 1995.

J. Vollmer, A. G. Vegh, C. Lange, and B. Eckhardt. Vortex formation by active agents as a model for daphnia swarming. *Physical Review E (Statistical, Nonlinear, and Soft Matter Physics)*, 73(6):061924–10, June 2006.

M. von Smoluchowski. Zur kinetischen theorie der Brownschen Molekularbewegung und der Suspensionen. *Annalen der Physik*, 326(14):756–780, 1906. ISSN 00033804. doi: 10.1002/andp.19063261405.

Wolfram Research, *Mathematica Edition: Version 7.0*. Wolfram Research, Inc., Champaign, Illinois (USA), 2008.

Acknowledgments

This dissertation would not be possible without the support and guidance of different people, which I would like to use the opportunity to acknowledge here.

First and foremost, I would like to express my gratitude to my supervisor Prof. Lutz Schimansky-Geier. His passion for statistical physics and stochastic processes was a continuous source of inspiration. I have strongly benefited from his deep knowledge and understanding of stochastic processes and our intense discussions on the mathematical and conceptual problems involved in my research were instructive and enlightening.

I am also deeply indebted to Prof. Iain D. Couzin from Princeton University, who I had the pleasure to collaborate with during my Ph.D.-research. His original and groundbreaking ideas on collective behavior, were – and still are – truly inspiring. His friendly advice and his remarks on biology of collective motion had a major impact on my work.

I would like to thank Sepideh Bazazi, a doctorate candidate from Oxford University during the time I was working on my thesis, for providing me experimental data on collective motion of desert locusts. I really enjoyed our friendly and successful collaboration which allowed me to push my research beyond pure theory.

Special thanks go to Dr. Udo Erdmann, who introduced me to Active Brownian Particles. His enthusiasm for the subject was infectious. Even after he left academia, his mentoring and support were invaluable.

Furthermore, I would like to thank all the members of the "Stochastic Processes" and "Statistical Physics & Nonlinear Dynamics" groups at the Department of Physics, Humboldt Universität zu Berlin. I really enjoyed the friendly, stimulating and productive research atmosphere. Special thanks go to Felix Müller, it was highly rewarding and enjoyable to work with him on our common side-project.

I must also acknowledge the help of Udo Erdmann, Simon Fugmann, Peter Kuras, Eliot Michelson and Momme Winkelnkemper with the revision of the manuscript.

I would like also thank my family and friends for the joyous moments and their support they provided me with throughout my entire life.

Last, but not least, I would like to thank the most important person in my life: Nina Seiferth. I cannot imagine how I would manage the past years without her deep love, continuous encouragement and unconditional support.